Grenada in Wartime:

The Tragic Loss of the *Island Queen* and other Memories of World War II

Grenada

in Wartime:

The Tragic Loss of the *Island Queen*
and other Memories of World War II

Beverley A. Steele

Copyright © Beverley A. Steele 2011
Presentation Edition 2011
First Edition 2011
Revised Edition 2013

Email: bevathome@spiceisle.com

All rights reserved. Except for use in review, no part of this publication may be reproduced or transmitted in any form or by any means, electronic or mechanical, including photocopy, recording, any information storage or retrieval system, or on the internet, without permission in writing from the publishers.

 Design and Layout by Paria Publishing Company Limited
www.pariapublishing.com

Maps by Xandra Fisher Shaw
Painting on back cover by Susan Mains
Typeset in Life
Printed by Lightning Source, U.S.A.

ISBN 978-976-8054-93-1

Also by Beverley A. Steele:
Grenada. A History of its People.
ISBN 0 333 93053 3
MACMILLAN CARIBBEAN 2003.

DEDICATION

This book is dedicated to the memory of those who were passengers on the *Island Queen*, and thereby lost their lives;

As a thanksgiving for the lives of all who wanted to travel on the *Island Queen* between Grenada and St. Vincent on that fateful weekend — August 5&6, 1944, and for different reasons, did not;

As a measure of closure and remembrance to the families and friends who suffered the irreparable loss of loved ones on that sad occasion

And to those too young to know or to remember the events of World War II in Grenada.

Lest we forget.

TABLE OF CONTENTS

DEDICATION	V
TABLE OF CONTENTS	VI – VIII
LIST OF MAPS	VIII
ACKNOWLEDGEMENTS	IX
INTRODUCTION	X – XIII

PART I: GRENADA'S RESPONSE TO WORLD WAR II

I	And So the War Began	1
II	Immediate War Measures	5
III	Tug of War for Grenadian Loyalty	7
IV	British War Propaganda	13
V	Caught Overseas	15
VI	Censorship	18
VII	Managing Scarcity	19
	The Competent Authority	19
	The Black Market	21
	The Food Supply	22
	Clothing	27
	School Supplies	30
	Motor Vehicles, Spare Parts, Tyres and Gasoline	32
	Luxury Items	33
VIII	Contributing to the War Effort	36
IX	Grenadians in Military Service	39
X	Spies	47
XI	News and Broadcasting	51
XII	Financial Hardship and Migration	55
XIII	The Children during Wartime	60

PART II: THE CARIBBEAN AS A THEATRE OF WAR

I	American Neutrality and the War	65
II	The Militarisation of Trinidad	69
III	The Militarisation of Grenada and other Eastern Caribbean Islands	73

PART III: GERMAN U-BOATS IN THE CARIBBEAN
I   The Rewards of Careful Planning   78
II   The Defenceless Caribbean   80
III   Achilles' Escapades   87
IV   Sirens and Blackouts   92
V   Attack on Bridgetown Harbour   95
VI   Other Ships Sunk near Grenada   96
VII   Rescuing Survivors   100
VIII What the Sea Brought in   104
IX   A Dangerous Harvest   106
X   The U-Boats Display Themselves   107
XI   Consequences Of U-Boat Activity in the Caribbean   112
XII   Schooners to the Rescue   115

PART IV: A SCRAMBLE TO DEFEND THE CARIBBEAN
I   Vessels for Defence   121
II   Aircraft   122
III   Zeppelins   124
IV   Motor Torpedo Boat Flotillas   125
V   Telecommunications   126
VI   The U-Boat Menace Lessens   126

PART V: THE STRANGE AND TRAGIC LOSS OF THE GRENADIAN AUXILIARY SCHOONER THE *"ISLAND QUEEN"*
I   Gordon Campbell's Excursion   131
II   Who were the Passengers?   133
III   Who were the Captains?   139
IV   The Schooners   141
V   The Voyage to St. Vincent   145
VI   Where is the *Island Queen*?   150
VII   The Wedding   151
VIII   The Search Begins   152
IX   A Sad Homecoming   154
X   Every Effort Made   155
XI   What the Newspapers Said   158
XII   Clutching at Straws   163
XIII   Official Closure And Messages of Condolence   166
XIV   Rachael Weeps for her Children   171
XV   Quirks of Fate   179
XVI   So What Really Happened to the *Island Queen*?   183
XVII   The Official Enquiry   192
XVIII No End to the *Island Queen* Story?   198

PART VI: THE WAR ENDS

I	Celebrating "V.E." Day	203
II	Results of War	205
III	The Mine Explosion in Windward, Carriacou	211
IV	The Grenada Veterans	214
V	Post War Development In Grenada	218

CONCLUSION: GRENADA TRIUMPHS!   221

BIBLIOGRAPHY   224

APPENDICES

Appendix I   232
A Listing of those who lost their lives on the ill-fated *Island Queen*, with brief biographical notes.

Appendix II   249
A Listing with brief biographical notes of those who were a part of the ill-fated Grenada to St. Vincent Excursion who travelled on the *Providence Mark*, and so saved their lives.

Appendix III   257
Reprint of the St. Bernard, Cosmo, *The Island Queen Disaster,* Grenadian Voice Newspaper Friday 30th July 1999.

INDEX   261

LIST OF MAPS:

Street Map of St. George's Grenada circa 1944
The Island of Grenada
Map of the Caribbean
The Windward Islands
The Islands of Trinidad and Tobago

ACKNOWLEDGEMENTS

First of all, I acknowledge that it is God who has given me the life and the ability to bring this work, so long a dream, to an eventual conclusion.

I am exceedingly grateful to my family and my friends who have exhibited generosity and patience with me while this book was in progress.

I am deeply thankful to Professor Edward Cox of Rice University who provided copies of relevant articles from *The West Indian* newspaper from which I worked. My grateful thanks also go to the staff of the archives of St. Vincent for allowing me access to their newspaper archives.

I am also indebted to Professor Sir Woodville Marshall who read versions of the manuscript as it evolved, and provided crucial, critical and continuing support and encouragement.

Without the information provided by over 50 respondents I would have not had the information and insight needed to produce this work. I am deeply thankful for their generous assistance. and the time spent with me sharing their knowledge and their precious memories

I acknowledge the special help of Shirley Charles, Gillian Glean Walker, Elinor Lashley, Getrude Simone Niles, Nellie Payne, Norma Sinclair and last, but certainly not least William Steele, who was a great resource for information on a variety of subjects.

I would like to acknowledge the generosity of Xandra Fisher Shaw who allowed me to use of her beautiful maps and artwork for use in this book.

I was unable to find a picture of the *Island Queen* other than at the boat's launching. Susan Mains filled this gap with a lively and imaginative painting of the *Island Queen* setting out on her fateful voyage to St. Vincent on 5th August, 1944. I am most grateful to her for her wonderful gift and contribution to the remembrance of this sad occasion.

Thanks to Paria Publishing for the design and layout of this book, especially to Alice Besson and Kelsea Mahabir. To those who lent me their photographs and to every single person who contributed in any way to the writing and production of this book, I will never forget you or your kindness.

INTRODUCTION

In the Harper Luxe 1993 edition of her book *To Kill a Mockingbird*, Harper Lee begged her readers to spare her an Introduction, because, she said, introductions inhibit pleasure, kill the joy of anticipation and frustrate curiosity. I wish I could simply state the same for the volume: *Grenada in Wartime: The Tragic Loss of the* Island Queen *and other Memories of World War II*. However, World War II enveloped the whole world in a way that the previous World War did not, nor has indeed any of the all too frequent subsequent wars. I am persuaded that persons who live outside of Grenada who were traumatised in one way or another by the events of World War II, or who maintain an interest in the happenings of this War, may want to read this volume.

This introduction is for them, and introduces not so much the history of World War II in Grenada, but locates Grenada for the reader unfamiliar with this Blessed Island geographically and historically. It also seeks to give the reader a glimpse of the Grenadian people, and importantly for this book, gives placement to this island as a society which found itself in the centre of the Caribbean Theatre of War. In this Theatre the War was almost lost to Hitler's Germany and her Allies.

Grenada is the most beautiful and the most southerly of the Windward Islands. It is a country of 133 square miles and a population today of about 100,000.Three islands make up the green and lush State of Grenada whose mountains reach for the sky, and whose rivers bound down these mountains, providing fresh water for all who need it.

If you need the exact location of Grenada, the line of 12 degrees north latitude passes through the southern end of the island, and the lines of 60 degrees and forty minutes west longitude and 60 degrees forty-five West longitude slice thought the island, the former almost through the middle. Trinidad is about 90 miles to the south, and St. Vincent about 76 miles to the north north-east. Grenada is 21 miles in length and 12 miles at its widest point. But Grenadians know that their islands are really located in their hearts, and the land space is just big enough — to live, to work, and to enjoy.

In the days when Spain, France and England were fighting for possession of Caribbean territory and at the same time were wresting these same lands from the native people who were the occupants and rightful owners, many battles were fought over Grenada. In the end, it was the British that walked away with this green gem. Grenada was continuously British since 1784 and until 1974 when she became an independent nation. Although Grenada's culture is a colourful mix of elements from all the ethnic groups that have made this island their home, British traditions still predominate. English is the official language, and Cricket and English Football compete for the number one favourite sport.

Grenada is a seafaring nation, and until the 1920s, boats built in Grenada and neighbouring islands provided the only facility for inter-island travel and commerce. These sloops and schooners of ancient design, aided by small modern "iron boats", still ply the traditional routes, providing alternative but important means of trade and transport. They are joined in this role by the aircraft and the most up-to-date shipping to link Grenada to its neighbours and to the wider world.

Grenada has two seasons – the wet from June to December, and the dry the other five months. In this country people have a habit of ignoring rules – doing things "to suit" – and this also applies to the weather and the seasons, which do not always conform to predictions.

At the time of writing Grenada is a fully independent tri-island nation. It follows a Westminster style of Government with a Prime Minister, Ministerial System, Cabinet, a Senate and a Governor General. The Governor General represents Queen Elizabeth II, who is the Head of State.

However, looking back to the time of World War II, Grenada was a colony of Great Britain and part of the Windward Islands grouping, which was under the jurisdiction of a Governor. Each island had an Administrator who saw to the day-to-day affairs, and who reported to the Governor. The Governor moved his location from time to time, living in rotation in each island under his jurisdiction. Each island had a Government House. Grenada was a Crown colony from 1815 to 1951 and during the 1940s, Grenada's Crown Colony Constitution was undergoing slow revision to allow some elected members and some nominated members to have a say in how the colony was governed.

Grenadians are a tough and resilient people, who have experienced enormous suffering throughout their history. What they cannot help, they accept as from God's hands, but they will rise up in arms if they believe that a wrong has been done that is in their power to right. This assertiveness was dearly won, as it was not rewarded under the plantation system, or under colonialism. Grenadians are also "Doubting Thomases", slow to believe anything, unless it is proved to them.

This quiet, developing small island, with its hardy population, was thrust unexpectedly into the 1939-1945 war, into which the whole world was drawn. If Germany had won the war, there would have been the loss of democracy and freedom. Grenadians were having none of that, and rose to all the challenges, including that of being in the middle of the Caribbean Theatre of War. But there is a price to each war – a price paid in the lives of young people and in the rape of the land. Britain had to face both, even as victors. Grenada was not left in ruins but, although no battles were fought on Grenada's soil, Grenada saw perhaps more action than some of the rural parts of Great Britain. She had her own territorial Army called the Grenada Volunteer Force and Reserve. She also sent young men and women off to Europe to assist Britain in winning the war.

Grenada also mourned her war dead, for three Grenadian young men did not come home from battle, and three young ladies died as a result of the war, two from Hitler's bombs and one from a fatal illness. Grenada had to cope as best she knew how with veterans, some of whom were no longer sound in mind or in body. Grenada witnessed the deadly effects of the U-boats, the shortages of war, and the terrors of seeing and hearing, but not being in full enough possession of the facts to be certain of what was happening in the seas and skies around Grenada. Around mid-night between 5th and 6th August 1944, 67 passengers and crew of the auxiliary schooner *Island Queen* disappeared during the 76 mile voyage between Grenada and St. Vincent. There is no proof as to the fate of the vessel and its occupants, allowing rumours, personal favourite theories and conjectures to persist. The disappearance of this vessel is, however, almost certainly war-related. This account will not attempt to solve the mystery of what happened to the *Island Queen*, but is offered as a memorial, both in terms of as complete an account as possible and in lieu of graves and headstones that could not be erected because no bodies of any of the passengers were ever found. This island mourned the tragic loss of her brightest sons and daughters then, and the loss is still mourned.

Therefore, when the definitive history of World War II is written, let not the sacrifices of the far-flung colonies be forgotten. For even in Grenada the scars remain after 70 years have passed. With both the triumphs and the tragedies of Grenada in the war kept in mind, therefore, let the story of World War II and its impact on Grenadians and Grenadian life begin.

This account of Grenadian experience during World War II is based upon available literature, but also fundamentally on the recollections of Grenadians who witnessed some of the events firsthand. Although every attempt has been made to double-check the facts, there will be some inaccuracies. Therefore, with the author's best intentions this book is neither complete nor completely accurate. With this in mind, the author invites the readers' involvement by pointing out omissions and corrections for inclusion in the next edition of this book.

BEVERLEY A. STEELE

GRENADA, OCTOBER 2011

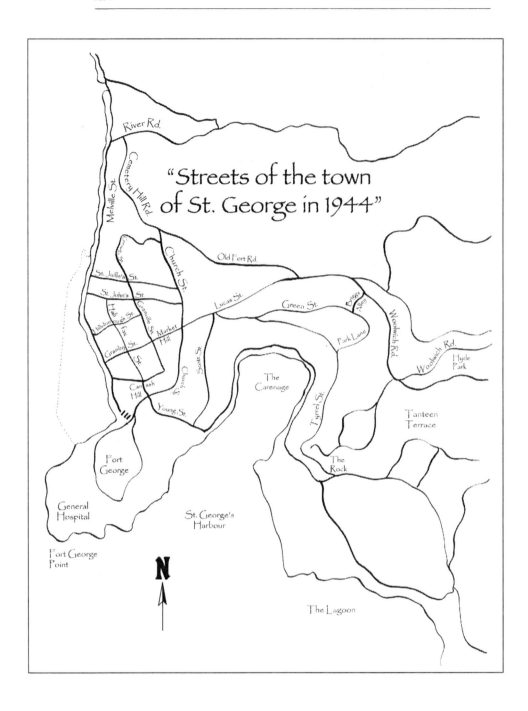

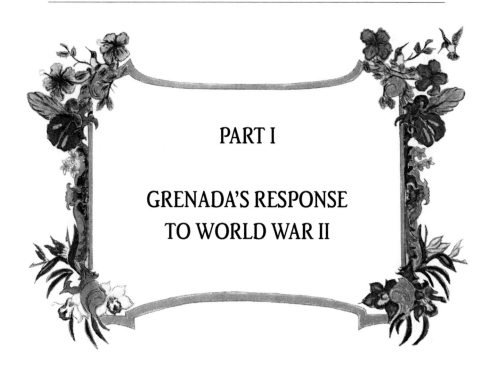

PART I

GRENADA'S RESPONSE TO WORLD WAR II

 I: AND SO THE WAR BEGAN

Imagine the scene. The German tourist liner *Columbus* with some 600 passengers preparing to enter St. George's harbour mid-morning on 1st September, 1939. Taxis have already positioned themselves at the waterfront, eagerly anticipating the disembarkation of the passengers who will hire them to go touring. The souvenir vendors are also abustle, hoping to get good sales for their carefully displayed items. To the amazement and consternation of the drivers, vendors and all onlookers, the liner, already very close to port, veers away, turns and steams out of view at full speed.

What could have happened?

Later it was learnt that the captain of the *Columbus* had received instructions not to go into the port of this British colony. If the ship docked, there was a strong possibility that she might have been impounded and that those of her passengers and crew who were German would be interned in Grenada as prisoners-of-war, for Germany had just invaded Poland, and World War II had begun in Europe. The *Columbus* and those on board came close to becoming Great Britain and Grenada's "first prize of hostilities." [1] Grenada's fledgling tourist trade has been profiting from the visits of the tourist boats to the island, many of which, including those of the well-known and popular Hamburg-American Line, were German. As German liners frequently arrived in St. George's, mainly to bringing tourists, these liners also carried passengers from Grenada to destinations in the United States and Europe. It was not surprising, therefore, there was another close call, this time for the German liner *Bremen*. This liner actually made it into Grenada's waters. She anchored in the outer harbour, but after a brief stay, was allowed to leave.

Although the taxi men, the craft sellers and other persons waiting to offer their services and wares to the tourists were nonplussed and disappointed, it was a good thing that none of these ships took the chance of being impounded in the St. George's harbour. If this had happened, there would have been serious implications for Grenada. Firstly, any ship impounded would have occupied scarce space in the harbour, possibly for the entire six-year duration of the Second World War. Secondly, the expense of impounding the ships and interning the German passengers and crew would have had to be met by Grenada. Everyone was lucky that these ships got away!

World War II had four beginnings. It began for Europe on 1st September with Germany's invasion of Poland. On 3rd September 1939, war was officially declared between Great Britain and France on the one hand, and Germany and Italy on the other. Two years later, in December 1941, Britain and her colonies would also be at war with Finland, Hungary, Romania and Japan. This declaration of war involved all of Britain's colonies. The beginning of the war for the United States would not be until December, 1941. It was from 3rd September, 1939, however, that Grenada would be at war with Germany and her Allies.

The possibility of war had been recognised for some time, and the serious-minded of Grenada's citizenry, who had access to a radio, had been listening regularly to the British Broadcasting Corporation (BBC) news broadcasts from London every morning at 7 a.m. and every evening at 7 p.m. to keep abreast of happenings in Europe. When Germany invaded

---

1   "Development of Tourism in Grenada" *Windward Islands Annual* 1962 P. 56

Poland, Neville Chamberlain, then Prime Minister of England [2], had made a recording to be played if his ultimatum to Adolf Hitler to withdraw his troops within three days was not met. The recording was broadcast as soon as the time for ultimatum passed. This broadcast was heard at 7 a.m. local Grenada time on that fateful day. In part it said:

> I am speaking to you from the Cabinet Room of No 10 Downing Street. This morning the British Ambassador in Germany handed the German Government a note stating that, unless they withdrew their troops from Poland by 11 a.m., a state of war exists between us. I have to tell you now that no such undertaking was received and, consequently, this country is at war with Germany. [3]

As soon as the news that war had been declared was received, it spread rapidly through Grenada, by various means. In St. George's, Grenville and Gouyave the church bells began to toll mournfully as if for a funeral. At this time there were few telephones, no widespread wireless service or radios in Grenada, and this was the best way to alert the population that something momentous was happening. Cosmos Cape, who was thirteen or fourteen years old at the time and who lived in the Grenville area, was an acolyte at the Roman Catholic Church in Grenville. The parish priest, Fr. Kelvin Clark O.P., asked him and another acolyte to take turns at tolling the bells for the whole day. He remembers being paid nine pence for his efforts.

Samuel Graham (later Sir Samuel) was the headmaster of Wesley Hall School when war was declared. Arnold Cruickshank remembers that, before school started on 3rd September, 1939, Mr. Graham explained to the pupils in language that they could understand the significance of the tolling of the bells, what war was, and how it might affect Grenada and the world. Adina Maitland [4] was an orphan in the Orphan Home situated near the Botanic Gardens. She remembers that Matron Augustine, a Barbadian, gathered the children together a few days before the declaration of war, told them that war was about to be declared, and said that when they heard the church bells start to ring they would know

---

2   In this book, the terms England, the United Kingdom and Great Britain will be used interchangeably. There may be a historic political difference in the terms, but they are all the same place.

3   Chamberlain, quoted in Murray, P. 15

4   Mrs. Irie Francis. The maiden names of all married ladies will be used throughout, with an appropriate footnote as to their name-change in latter life.

that the war had started. When the bells did start to toll, the Matron cried out in distress, and gathered all the children to pray. From then on the orphans prayed every day for all those fighting in the war.

Alister McIntyre (later Sir Alister) lived in Gouyave at the time war was declared, although he moved with his family to St. John's Street in St. George's the following year. He remembers the church bells of Gouyave ringing to announce the war. Norris Mitchell and Hermione Greasley [5], also from Gouyave, also remember distinctly the pealing of all church bells from the churches in Gouyave. When war was declared, Bertrand Pitt lived in Grenville with his parents. His father, Cyril S. L. Pitt, was the Chairman of the St. Andrew's District Board. He remembers that in a very short time the news of the war's beginning had spread through the length and breadth of the town of Grenville and beyond. Even before the bells started to peal, some had heard the news by radio, telephone and word of mouth. Janice Bain's [6] mother Kathleen received a telephone call from her father Bert Bain who was already at work, to tell her the news. She remembers clearly her mother's cry of dismay.

During the day there was much discussion on the war, the possible consequences for Grenada, and speculation as to how long it would last. Most people thought that the war would not last longer than the First World War or "The Great War", as it was sometimes called, which lasted from 1914–1918. Women with sons at or near the ages when they could join the armed forces began immediately to fret about the possibility of them going off to fight the war.

The very day war was announced, the Legislative Council met in emergency session. The first item on the agenda was to dispatch a message to Secretary of State for the Colonies expressing loyalty and fealty to the King, and pledging support. The dispatch read:

> At this momentous hour when the Armed Forces of the Empire have once again been called upon to uphold the cause of righteousness, the people of the Colony of Grenada, through their chosen representatives on the Legislative Council, with their humble duty to Your Majesty, desire to affirm their dutiful and abiding loyalty to Your Majesty's

---

5   Hermione first married Justice Barrymore "Barry" Renwick. After his death, she married George Hanson.

6   Mrs. Oliver Harbin

> Throne and Person and unreservedly to offer Your Majesty their services in any capacity which may appear to Your Majesty's Ministers in the United Kingdom helpful to our common cause.[7]

The Acting Governor also immediately prepared a statement which introduced and incorporated this message for the next issue of the *Government Gazette* and for the local newspapers.

The agenda for the meeting of the Legislative Council included the calling out of the Grenada Volunteer Force to man vulnerable points in Grenada. With immediate effect watches began at the wireless station at Old Fort, the Cable Hut at Queen's Park, the Power Station which was then at Burns Point, and at the gasoline reservoirs or "bowsers" that were then situated at the end of the Carenage, near to the present waterfront restaurants. Later there was an attempt to camouflage these bowsers by hanging burlap "crocus" bags from the protective railing around the tanks.

Finally, the Legislative Council decided that a public meeting should be held in the St. George's market square, at 4 o'clock that very afternoon. This public meeting took place as scheduled. The crowd that turned up was addressed by the Acting Governor and Mr. Charles F.P. Renwick, a prominent lawyer and member of the Legislative Council. The speakers reiterated the news that Grenada was at war, and conveyed to the public the immediate implications of this.

II: IMMEDIATE WAR MEASURES

That there would be a war with Germany had been expected, and England had advanced and detailed preparations in hand. Two weeks before war was declared, on 26th August 1939, the Ag. Governor of the Windward Islands had received an urgent dispatch from Britain informing him of an Imperial Order-in-Council, which extended the provisions of the Imperial Emergency Powers Act to Grenada. This Act was to enable swift Government action in Grenada once war was declared. Aided by this, it

---

7   Quoted from *The Grenada Handbook and Directory* 1946, P. 363.

was immediately possible to pass the Colonial Defence Regulations, to constitute the Office of the Competent Authority, to set up a system of price control in Grenada, to establish telegraph and postal censorship, and to form a War Purposes Committee.

In the days to come, and one after another, other Ordinances were drafted in England and transmitted to Grenada and the other British Colonies. In turn, these were passed by Grenada's Legislative Council. One of these was The Trading with the Enemy Ordinance [8]. Although Grenada's traditional trading partners were Britain, Canada and the United States of America (USA [9]), there was considerable commercial activity with Germany. Grenada imported engines from Germany, and persons with interests in Grenada also used Germany for services such as printing postcards of Grenada. As has been previously mentioned, vessels of the Hamburg-America line were frequent callers to this island. All such dealings with Germany were immediately suspended.

The acting Governor of the Windward Islands at the beginning of the war was His Hon. A. A. Wright. He was in office until 6th October, 1939, when the substantive Governor, Sir Henry Bradshaw Popham, K.C.M.G., M.B.E., resumed duties. Sir Arthur Francis Grimble K.C.M.G., M.A. replaced Popham on 18th May, 1942, and served until 1948. It was the job of these two Governors, particularly Grimble, to attend to everyday matters in Grenada, and to do what was necessary to protect Grenada, not only from the enemy attack by sea or air, but also from the enemy who might attack from within through espionage and information gathering. It was also the task of the Governors to maintain morale in Grenada, and peace among its people.

Grenada had a municipal system at this time, and each parish had a District Board comprised of both elected and nominated members. The District Boards had the responsibility of looking after the welfare of the parish under their authority, and the Chairmen performed a role almost like that of a mayor. The Chairman of the District Boards also had new roles to play. The Chairman of the District Board for St. George's during the war was Arnold Williamson.

---

8   Ordinance No 19 of 1939

9   The United States of America will be referred to also as the United States, the US or the USA or America throughout this book. This is by common usage, although the term "America" properly refers to both North and South America. Where North, Central and South America are meant, this will be so designated, or the term "Americas" will be used.

III: TUG-OF-WAR FOR GRENADIAN LOYALTY

What sort of attitudes prevailed in Grenada when Britain entered World War II? In 1939 Grenada was a society in transition from a plantation society into a modern society. It was still suffering from the demise of the plantation economy, and the amenities of modern life were enjoyed only by the small middle class and by those living in the capital city. The majority of people were very poor and just managed to provide themselves and their children with a bare and basic living.

Moreover, the planters and those persons in the society who identified with this class still clung to pre-emancipation attitudes of the plantation society, which included a lack of compassion for the very real poverty of the working class and an abhorrence of the idea of social equality. Workers were paid wages that they could not survive on, leaving them to make up the shortfall however they could. Long hours, back-breaking work and abysmal conditions of work were customary for field labour and house servants alike. In the 1920s there were strikes in Jamaica, British Honduras, Grenada and in other parts of the British Caribbean, The fledgling labour movement in the Caribbean with such leaders as Grenada-born Uriah "Buzz" Butler, Trinidadian Arthur Andrew Cipriani and Jamaican Alexander Bustamante had begun. Ten years later, there were serious disturbances in Trinidad, British Guiana, St. Lucia, Barbados, and Jamaica. The general militancy of workers, fuelled by resentment bred of oppression and deprivation, low wages, unemployment, and the imbalance of the spread of wealth, and triggered by such events as the attempt to suppress the development of the Trade Union movement and its leaders, threatened to disrupt the societies of the entire British Caribbean. Although labour unrest was slow in coming to Grenada, and there would be no real change in labour relations until after the war. The authorities were aware that in all islands there was the potential for unrest, and this was a major concern and matter for careful and continued vigilance during the war.

They need not have feared too much for the loyalty of Grenada. Any bitterness related to their circumstances in the mass of the Grenadian population was not directed at England or the King. It was directed at the Grenada elite. The declaration of war unleashed a flood of patriotic feelings and sympathy for England at the ordeal which she now faced, and most were prepared to shelve their aspirations for better wages and a more just society and come out in full support of England for the duration of the war.

How was a population, the majority of whom were descendants of enslaved people, transformed into loyal subjects of the King of England in just 100 years? Even before emancipation, Britain began to aim at achieving social stability in her colonial empire in the Caribbean through socialisation, and the integration of soon-to-be new citizens into British culture through the pervasive efforts of several agents. Familiarity with British culture was coupled with a sense of belonging and identity as subjects of the British Empire. Thus at emancipation the newly-freed population believed that their freedom had been personally given to them by Queen Victoria, and had been due only to a lesser extent to the efforts of William Wilberforce. The names of Thomas Clarkson and Granville Sharp, Hannah More and Charles Middleton and of others who were very active in the Emancipation Movement in England were virtually unknown in the Caribbean. In the post-emancipation era, generations of Grenadians grew up in a colonial culture that portrayed Britain as powerful, paternal and just. They came to firmly believe that it was their good fortune to be a part of the British realm and to be "Her/His Majesty's Subjects," and that the English monarch was personally responsible for the welfare of Grenadians, which was sometimes interfered with by the planters.

Long before the beginning of World War I, Grenadians were loyal to the Crown and belonged to the "Cult of Empire".[10] Insofar as it was possible, given their economic circumstances, they were more "British than the British". England was for the majority of Grenadians a place of overwhelming sentimental importance. The monarchy and its institutions were revered, and the laws and officers of the Empire respected. The *Union Jack* was the beloved flag of Grenada, and was handled with reverence whenever it was to be flown. The Church of England [11], although second to the Roman Catholic Church in terms of numbers in Grenada, was the church with the most prestige, and was honoured as such. The King or Queen's Birthday was

---

10   Metzgen and Graham, P. 107

11   At this time the Anglican Church was still called The Church of England and the Anglican High School the Church of England High School for Girls. These older names will be used throughout this work.

remembered and celebrated each year, especially by school children. Grenada also enjoyed the same sports as the people of Britain, such as cricket and netball.

Although a "poor relation" in terms of resources, Grenada had a similar educational system as Britain, with almost the same curriculum content and text books. The curriculum for primary schools and the few secondary schools was based on the curriculum used in England. English geography, English history, and English literature were taught to Grenadian schoolchildren, who were encouraged to think of themselves as English. Ulric Cross, a Trinidadian, could be speaking for any middle-class Grenadian and some of the working class when he describes how exposed he was to Britain and British sentiment at this time:

> To some extent I was not entirely unfamiliar with England, although I had never been to the country. My uncle was a doctor in Bolton, Lancashire...ever since I was a child... *The Overseas Daily Mirror, London Illustrative News, The Trailer (sic)*, a number of British magazines; we got regularly every month from him. One got, through these magazines, a very good insight into a cross-section of English life. We also knew quite a lot about the English at school. I knew more about the history of England than the history of the West Indies ...[12]

In addition to learning about England and English culture, gestures of loyalty to the British Empire were a daily occurrence in the school and in the classroom. Every child learnt the British National Anthem early and sang it whenever appropriate. *Rule Britannia, There'll Always be an England* and other British patriotic songs were also well known and sung on occasion, and with pride and gusto.

Just before World War II broke out, Grenadians were thrilled to be involved in the celebrations for the coronation of King George VI and Queen Elizabeth. Many people obtained pictures of the new King and Queen, and put these up in their homes, and the little princesses, Elizabeth and Margaret, were loved, admired and emulated by Grenadian children.

Magazines and newspapers from England were received by the better-off Grenadians, and passed down to their servants and estate workers. These printed pages were the source from which poorer people got

---

12 Murray, P. 19. *The Trailer* is possibly *The Tatler*.

treasured pictures of the Royal Family for their walls. BBC broadcasts were transmitted and received in Grenada and in most of the other British colonies. When the estate owners or other of the more affluent Grenadians listened to their wireless sets, they might also allow their servants to listen.

In all these ways English culture trickled down through the society. Because of their English colonial socialisation, most Grenadians at the declaration of World War II were therefore kindly disposed towards the Mother Country. This was in spite of the extreme poverty experienced by estate and other workers, the lack of health care and other amenities. Those who realised that the conditions of the working class in Grenada could be improved also had the knowledge that Grenada was on the brink of amelioration of some of the abysmal social and economic conditions. The Royal Commission, commonly called *The Moyne Commission*, had visited Grenada in 1937 to collect evidence on every aspect of Grenadian life. The commission had not only taken evidence from a cross-section of each colony, but had visited several places in each to get a "feel" of the circumstances for themselves. Grenadians believed that England would act on their findings and recommendations, and their belief was not in vain. Although many projects and progress towards modernisation would have to be shelved due to the onset of the war, the promise of reform would suffice while Britain fought for her existence and for the existence of her Empire. Moreover, although the *Report of The Moyne Commission* was not released until after the war, as the findings were very critical of the state of the colonies covered, some of the recommendations were acted on before the report was released, including major reforms in education and health, and developments in communications.

Loyalty to Britain was one issue, but another was the understanding of the implications for Grenadians of German totalitarianism and the Nazi doctrine of Aryan superiority. Neither of those doctrines promoted by the Third Reich appealed to Grenada's individualistic people. The prospect of having to limit their individual freedom and individual thought did not appeal to a population who were pressing for enhanced representative Government, more freedom and wider horizons. The tenets of German totalitarianism were exactly the opposite of Grenadian aspirations, and Hitler's doctrine of the superiority of the "Aryan race" was absolutely unacceptable to a population of largely African descent in Grenada, whose ideal was racial equality. These were the reasons why the majority almost unanimously put aside any disagreements and grievances and with few exceptions directed their energies as resistance to another enemy. When Britain declared war on Germany, the war was seen as their cause by the people of Grenada, and Grenadians rallied round the Union Jack.

There were, however, a minority who were anti-war and whose anti-war sentiments were bolstered by propaganda put out by the Negro Socialist Movement in the USA that found its way to Grenada. Resistance to fighting for the Allies was promoted first on the grounds that Negro soldiers would not be fighting for the rights of small nations and individual democratic rights. A surprising number of Caribbean people knew about, and were incensed at, the inaction of the League of Nations in Ethiopia, at this time called by the older name of Abyssinia.[13] Abyssinia was left unsupported and unprotected and its Emperor vulnerable to the assault of Benito Mussolini's Italian Army in 1936. As a result, that country was overrun, plundered, and Haile Selassie, the Emperor, had to flee for his life. The case of Ethiopia described in the socialist pamphlets was therefore taken to heart.[14] The cases of South Africa and Kenya were cited and found sympathy as well as that of Ethiopia as instances where the Allies had not in the recent past used force in support of non-white countries and native black populations. Secondly, the anti-war and anti-involvement pamphlets argued that Negroes needed peace, not war, for organisation, education and struggle against oppression, but the white man wanted war because war was profitable. It was also mooted that only by the abolition of imperialism and of the whole network of war-makers could peace be achieved. With peace would come the ultimate freedom of colonial people.

Some pamphlets detailed the experiences of Black American soldiers in World War I, and asked if Negroes should again expose themselves to similar experiences. The pamphlets told of how in the USA, Black soldiers had been segregated in "Jim Crow" Regiments, which were deployed as Labour Battalions for loading and unloading ships, building roads and depots, digging latrines, burying the dead, and detonating scattered

---

13 Claudia Jones and J. R. Johnson both describe the effect the reluctance of the Allies to go to the assistance of Abyssinia had on Black People in the Caribbean and America.

14 To the delight of Caribbean people, on 5th May, 1941 Haile Selassie returned to Addis Ababa after Ethiopian tribesmen were, in the end, joined by British Imperial Forces. The combined forces routed the Italian army from North Africa.

explosives. Black soldiers were used as servants, cooks and clean-up crews, even though they had enlisted as soldiers-of-line. Negro soldiers had been jeered at, tormented, and five had been lynched in uniform. Grenadians could relate to these sentiments, as Grenadian soldiers serving in World War I had met with open prejudice, disrespect, and assignment to the most menial jobs. It is undeniable that the attitudes of the military authorities in Britain and the USA towards black soldiers had been appalling.

In the face of these prejudices and experiences, the question in some minds was: should Grenadians expose themselves to this treatment again? Should Grenadians fight a white man's war? The "race" questions within the military and in the British and American society handed Germany a powerful propaganda weapon. German agents plugged this unfortunate state of affairs for all it was worth. The anti-war propaganda was so effective in that some recruits for the Armed Services were persuaded to quit, sometimes at the last moment after completing their training. In Grenada, distinguished citizens loyal to the Crown tried to neutralise the propaganda as best they could. In a speech to the St. Andrew's Detachment of the Grenada Contingent, C.H. Lucas, a respected coloured lawyer and politician, expressed the following sentiments:

> In spite of the mischievous activities of a few, happily very few; in spite of the ignorance of some; in spite of the arrant cowardice of others; over fifty of our fellow parishioners are here today to attest their patriotism and their loyalty. I would remind those who would be seditious that their freedom of lying pro German talk is only possible under the generous rule of Britain. The malevolent talk about being "windbreaks" for the British troops is meant to injure recruiting, but only deserves the ridicule of those who know the conditions of modern warfare.... As to the cowards who first signed on, then were medically examined, and at last turned tail and ran away from the oaths — well, I will not parley with cowards.[15]

If they could not be loyal out of conviction, many realised that to behave otherwise was not in the individual's best interest. Members of the Police Force were instructed to look out for the few dissenters and to deal with them as appropriate.

---

15   Lucas, P.1

IV: BRITISH WAR PROPAGANDA

To sustain traditional loyalties during the hardships of war, Britain took deliberate measures to ensure support for her war efforts and measures in Britain and her colonies. A document entitled *Principles Underlying British War Time Propaganda 1939* was written in Britain and endorsed by the British Prime Minister, Neville Chamberlain. It said in part that Britain must design and disseminate propaganda that was capable of

> Rousing the whole nation and inspiring them with devotion
> to the national cause. It must be able to represent that cause
> as a crusade, for which no sacrifice is too great. [16]

This *esprit de corps* would be communicated to the people of Britain and the British Empire by various means such as speeches, news services, topical comment, films, broadcasts and photographs. In essence, the message would be that Britain was fighting for not only her safety and integrity and that of the Empire, but for the preservation of Western civilisation. The war was for the continuance of the Christian ethic, the scientific spirit, and the rule of law, absolute values, the importance of the family, the sanctity of the individual rights and freedoms, and of international rights. The assertion was that Britain would win the war because of her great strength. However, the maintenance of this great strength called for correspondingly great sacrifice. Each member of the British Empire was called to play his or her part to ensure victory.

The efforts at this sort of propaganda were not in vain, for Grenadians in general were thrilled at the speeches of Winston Churchill and others, who became their heroes. Pictures and posters were pasted everywhere. Grenadians took in stride the occasional ignorance displayed by the persons in charge of disseminating visual war propaganda. In Grenada there was a poster whose graphics were designed to show the size of a tank. The poster showed the tank next to an elephant. How many Grenadians had experience with elephants? Alphabet books in Grenada predictably cited "E" for elephant. There were also travelling circuses which visited Grenada occasionally complete with lions, tigers and at least one elephant. Many Grenadians walked for miles just to get a

16   PREM 1/441

glimpse of these animals behind the fence. Nevertheless, elephants were definitely not common to the experience of the average Grenadian.

Free books were also given to children on aspects of the war. Dunbar Steele remembers some of these featuring *Spitfires* and *Hurricanes* and other war machines such as tanks that would appeal to the imagination of schoolboys, making them enthusiastic about the war effort.

From the very start of the war there was a blanket of secrecy about what transpired in the sea that lapped the islands. This secrecy and the lack of war experiences allowed Grenadians to carry on living as if the war did not exist. War seemed unreal and beyond the imagination of many, especially as the news that came through was news of distant and foreign theatres of war. Most Grenadians never realised the seriousness of the war in the Caribbean, that they were almost entirely unprotected, and that Germany could have captured Grenada and most of the islands of the Eastern Caribbean at any time between 1942 and 1943 if they had cared to do so.

What was very real to Grenadians was the encroachment on their lifestyle. The Secretary of State for the Colonies fully realised the importance of this for Grenadians as well as for all people of the British Empire. The text of a Circular Telegram to all Governors in part read:

> It must be appreciated that, as financial strain of prolonged war on Empire Resources is likely to be great, all expenditure by Government (and indeed by general public) should, as far as possible, be avoided which a) involves use of foreign exchange...b) creates demand for unessential goods and so deflects men, material and shipping from war purposes. In this connection you will appreciate that it will be very desirable to replace imported goods of all kinds, especially foodstuffs, by local produce wherever possible.... I am anxious to see existing social services and development activities disturbed as little as possible both because retrenchment and serious curtailment of services at present juncture might have very unfortunate effect on Colonial people and also because on grounds of policy it is important to maintain our reputation for enlightened Colonial Administration. In particular I am anxious to avoid any retrenchment of personnel. [17]

In one breath, the Secretary of State was curtailing all spending, the use of foreign exchange, and access to manufactured goods and access

---

17 CO 321/386. Dispatches. Windward Islands Grenada 1939. Circular Telegram from Secretary of State for the Colonies to all Colonial Dependencies except Palestine and Trans-Jordan. d/d 15/9/1939

to shipping. At the same time he was advocating import substitution in colonies historically designed to be dependent on the Mother Country for manufactured goods and discouraged from being self-reliant. Then he expected the Governments to protect local jobs and maintain social services and development activities! He must have known he was asking the impossible. However, Grenadians have a saying: "The difficult we do right away, the impossible takes a little longer."

V: CAUGHT OVERSEAS

Those Grenadians who had family members overseas when war was declared were in a state of acute anxiety. In most cases all efforts were made to have family members overseas return safe and sound to Grenada. The beginning of the war caught Constance Williamson in New York. Constance was the wife of Grenada's prominent citizen and chairman of the District Board of St. George's, Arnold Williamson. Efforts to secure an immediate passage to Grenada were frustrating, because these were virtually impossible to get. Eventually after much trouble and worry she returned safely to her very anxious family.

Those families who sent for their children who were studying in England had to accept long and fretful passages for them routed through Canada or America and then on to the Caribbean. Although no Grenadians travelling home were actually on ships that were torpedoed, some of their friends were. Eileen McIntyre [18] records that the ship on which one of her young friends was returning to Trinidad was torpedoed, and the passengers had to spend many hours in a lifeboat until they were rescued. Some English post-graduate students attending the Imperial College of Tropical Agriculture in Trinidad lost their lives when the boat on which they were returning home to England was torpedoed.

It was the choice of other families with children studying at university in England that it was best to keep their young people there to complete their studies. Some of these students joined the armed services in England. One of them was Noble Hudson, a Grenada Island Scholar, who was in England studying medicine. When World War II broke out, he joined the Medical Corps of the Royal Air Force.

When a child remained abroad for study or to live, the family left in Grenada had to find a way to deal with the daily anxiety. At the time war was declared, Shirley Buckmire [19] was boarding with the Jacobs family on Lucas Street in St. George's. Their daughter Gladys [20] was in England, and the family was in constant anxiety for her safety, especially when there was news of bombings and the Blitz.

Most of the families of young people who were about to leave to further their studies decided not to send them at all. In the end, some lost their opportunity to study altogether. One such young man was Edward Kent, who was scheduled to depart for England to study Law. Because of the dangers of trans-Atlantic travel during the war, his father kept him in Grenada, and sent him for the interim to work on his estate in Carriacou. By the time the war was over, Edward had become wedded to agriculture, and no longer wished to tackle the legal studies.

Also anxious to get home as soon as possible were those who were visiting Grenada at this time. A number of years previously, Christine Gun-Munro's [21] father had studied medicine in England, and stayed to practice there. He and his wife were visiting family in Grenada in 1939 when war was declared. They immediately took the first available boat back to England — a boat hardly better than a "cattle boat", with very rough accommodations. The boat travelled in the irregular zig-zag path advised for all shipping as a way to foil U-boat [22] attacks, as U-boats were active in the Atlantic from the first day of the war. However uncomfortable the passage, this was a fortuitous decision because the liner on which the Gun-Munros had originally booked was torpedoed in the Atlantic by a German U-boat. Some passengers were rescued from the sea, but there was also some loss of life.

The Gun-Munro family escaped the fate of Rosamond and John Barker-Hahlo. Rosamond Harford, originally from Vendôme, had migrated to Trinidad in search of suitable work. She married an Englishman and retired naval officer, John Barker-Hahlo, who was working for Trinidad

---

19   Mrs. Rawle Charles

20   Gladys later married Sir Arthur Lewis.

21   Mrs. Gordon Brathwaite

22   Short for the German "Unterseeboot", a popular way of referring to a German submarine, which will be used throughout this book.

Leaseholds Ltd., an oil company at Pointe-a-Pierre. The wedding was in August 1939, and the couple set sail on the *Regents Tiger* intending that the voyage would be their honeymoon before they began their new life in England. War was declared while the couple was on the high seas, and their ship was torpedoed by a German U-boat in the Bay of Biscay off a point called "The Lizard" on England's south coast. Thankfully, Rosamond and John, along with some other passengers and crew, were rescued by a Belgian ship, but they lost everything they had. As soon as he could, John Barker-Hahlo re-enlisted in the Royal Navy and was re-commissioned as a Lieutenant Commander. John Barker-Hahlo was killed in action two years later when his ship, the *H.M.S. Southampton*, was dive-bombed and sunk in the Mediterranean. He had never seen nor held his infant son, George Robert, although he knew him through the letters and pictures sent to him by his wife.

When Germany overran France in 1940, Guadeloupe and Martinique, France's colonies in the Caribbean, were also under German occupation. Eric Glean, then a young man of 31, was working with the Ford Motor company in Guadeloupe. He had no intention of remaining in a Nazi-occupied island and, using his well-known talents as boat-builder and sailor, he devised a plan to 'escape'. He built a sailing boat, which he named *QED*. He then took *QED* sailing every weekend, until the coast-guard grew accustomed to seeing 'Monsieur Glean' in his little sail boat, and therefore paid no attention to him as he sailed south through Guadeloupe's Les Saintes islands, and sailed back again. Then one afternoon, Eric sailed past Les Saintes and on to Dominica and Martinique until he reached St. Lucia. There he was able to recruit a crew of two. These were Compton from Carriacou and Ben Roberts, who was to serve later in his life as Grenada's Commissioner of Prisons, for the longer sail home to Grenada. It was not until Eric Glean failed to show up to work on Monday that enquiries revealed that the coast-guard had not seen him return from his usual sail!

Unlike trans-Atlantic voyages, travel around the Caribbean was quite safe up to the advent of the U-boats in the Caribbean. David Otway was on holiday in Trinidad when the war broke out. He came home on one of the Lady Boats. Students at secondary school, teacher's college, or other institutions of higher learning travelled as usual to Trinidad and Barbados usually by schooner or sloop, and returned to Grenada for the holidays. Persons working in Trinidad came home regularly to see their families. The return fare to Trinidad by schooner was ten dollars.

The Governor of the Windward Islands also had to travel by sea from Grenada to the other islands under his jurisdiction. He sometimes travelled with his wife and support staff. Other officials such as judges and very senior civil servants needed to travel from island to island in fulfilment of their duties. Even after the U-boats made their appearance in the Caribbean, inter-island travel continued, because the U-boats would seldom challenge the sloops and schooners as they plied their path between the islands. They were sent to the Caribbean after much bigger prizes than a schooner or sloop filled with island people. People were willing to take the risk of sailing through a sea shared with the U-boats so that they could to get where they wanted to go to do what they had to do.

VI: CENSORSHIP

As a result of the war, telegraph and postal censorship was established for the eastern and southern Caribbean, with Trinidad as the headquarters. Grenada's censorship activities were under Commander Kassell, the British Commander in charge of all aspects of war in Grenada. He was ably assisted by a number of Grenadians, including Frieda Martin, Laurie Harbin, Mable Gibbs, Mollie McIntyre, Estelle Garraway, and Hugh Henry Pilgrim.

Censorship required that a return address be written on all overseas mail, otherwise the letter would not be accepted for posting. A letter would be returned to the sender if it contained the type of information that was not approved. Letters coming into the island from abroad were carefully censored, and those that had information in them that was considered dangerous to security had the relevant sections cut out. Letters stamped CENSORED are well remembered by many. Incoming cables were also scrutinised by censors at the cable office before they were delivered, and each outgoing cable was vetted. One of the censors at the cable office was James Brathwaite, a former Government Auditor. Sensitive cables were received in code, and were deciphered by a cadre of very trusted citizens, including Kathleen Gibbs.

VII: MANAGING SCARCITY

Grenadian patriotism at the beginning of the war was flavoured with anxiety and the certain knowledge that there would be profound changes in the lifestyle of Grenadians. In 1939, apart from the products of a few small home industries and the production of raw sugar, nothing was manufactured in Grenada, and a good deal of Grenada's food was imported, as were most of the items used daily. The supply of goods from overseas was dependent entirely on sea transport from Canada, the United States, and Europe. Although merchants immediately sent additional orders for such things as flour, soap, salt meat, salt fish, onions, English potatoes, and powdered and condensed milk, certain types of food were soon unavailable or only available periodically. In only a short space of time, to find a tin of condensed milk, sardines or corned beef on a grocery shelf, was like being awarded a prize. This was because there was an almost immediate curtailment of shipping due to the operation of the U-boats in the mid and North Atlantic from the first day of the war. The U-boats sent tons of shipping with goods meant for the Caribbean and other parts of the world to the bottom of the sea. The situation worsened when the U-boats were unleashed in the Caribbean area at the beginning of 1942.

The Competent Authority

Because there would be shortages of all goods imported into Grenada, a body known as the Competent Authority was established with a mandate to take whatever action was necessary to manage available resources and to stabilise the economy. This Authority, headed by Gittens Knight, a lawyer and otherwise very well read, took immediate control over all imports, especially essential items. Licenses had to be obtained from the Competent Authority for the importation of all goods. A schedule was put in effect that listed what items could be imported, and by whom. Certain items were scheduled and more closely regulated than others, and at the top of the list there were some items that could only be imported by the Controller of Supplies.

The Competent Authority also had the power to regulate the price of everything that was offered for sale. At the outbreak of the previous

World War, there had been an immediate upward revision in the price of goods by the merchants, causing distress to the population. This was not to happen again. In addition to price control, the Competent Authority had the right to subsidise the cost of some goods, and although the cost of living would rise, it did not rise as high or as immediately as it would have without the imposed regulations.

The Competent Authority used a form of rationing to control the distribution of essential goods. Among the items rationed were gasoline, kerosene, other petroleum products and tyres for both motor vehicles and bicycles. The shortage of tyres became even more acute when in February 1942 the Japanese seized Malaya and Singapore, the major producers of rubber in the world. There was also some rationing of food items, which generally took the form of allocating certain commodities only to specific agencies and traders. There were strict rules in force as to whom "scheduled items" were to be sold, and the quantity which was to be sold to each customer. Moreover, householders were not permitted to have more than one week's supply of anything, and shopkeepers were not permitted to hoard items. Traders and shopkeepers had their premises inspected periodically to ensure that they sold what they had, and did not hold back goods against the hope of a price rise. The traders had to submit a return to the Competent Authority on any of their stock of "scheduled items". Motor vehicles, tyres, spare parts, gasoline, kerosene, rubber, leather, salt and matches were all "scheduled items".

The most important food commodity rationed, as far as Grenadians were concerned, was wheat flour. When this was to be had at all, it was consigned exclusively to bakeries such as *Haynes Bakery* on Jack's Alley, *Sardine's* on the Carenage, and *Maggie Mitchell's Bakery* on Melville Street close to the abattoir, all in St. George's, and to others scattered across Grenada. Even then, the supply was far short of the demand. When bread was available for sale, news would travel quickly and there was much pushing and shoving and the occasional fight to buy a loaf or two. Alister McIntyre, on hearing that bread was available at *Maggie Mitchell's*, decided that he was going to get some for his family. He had never before seen such a scene of pushing and shoving as he did when he approached the shop. The only reason he was able to get some loaves is that he was already well known and well liked around town, and men in the crowd picked him up and put him on their shoulders, from which perch he was able to transact the business at hand. Frederick McDermott Coard, in his autobiography written many years after the war, thought it important to mention that an invitation to tea by the Chief Revenue Officer during the War was remarkable because "I saw some welcome slices of bread on the table, a sight I had not beheld for several weeks". [23]

It was decided that His Majesty's Prisons at Richmond Hill would be allocated enough flour to bake bread to feed the inmates of the prisons as well as people in the Government institutions. Each day the bread from the Prisons Bakery was brought to the various institutions — including the hospital, the Mental Hospital and the Orphanage — in a bread push cart. The bakery at the prisons gained a reputation for baking excellent bread. The tradition that started as a war measure continues, and the prisons continue to bake superb bread, unfortunately never on sale to the public.

The Black Market

As a reaction to the shortages and the imposition of restrictions on imports, an informal trade developed to import goods, mostly "scheduled items" and those not on the priority list of imports, into Grenada by small boats, usually from Trinidad. These items were sold privately or shared by the captains of the boats among family and friends. Those who traded informally were subjected to heavy fines if they were caught, but perpetrators were seldom hunted down and prosecuted because illicit trading was an important way in which the population got by. The most popular import by schooner was wheat flour. Even traders who imported other goods legally by schooners and sloops would also instruct the captain to bring in flour "on the QT", i.e. quietly. The flour would then be sold among his family, his staff and his friends. One little girl, the daughter of a St. George's merchant who could get as much flour as he wanted, often went to school with mouth-watering white flour "bakes" — a large fried dumpling also known as *Johnny Bake* or *Journey Cake*. She was the envy of her classmates, who would look at her lunch hungrily, and beg her for a piece. In no time at all she acquired the nickname "Bakes", by which she was known for a long, long time.

Gasoline, leather, tyres and other goods were also brought in and sold informally, outside the aegis of the Competent Authority.

---

23   Coard, P. 90.

The Food Supply

The shortage of imported food forced adjustments to what one ate. The adjustment to the diet is one of the best remembered impacts of the war on Grenadian life. There was an Official *Grow More Food* campaign for England as well as for all the British colonies. Grenadians were urged, and did heed the call, to grow more food and rear more animals in order to meet the exigencies of war, to replace unavailable imports and to avoid starvation.

The Agricultural Superintendent in Grenada during the war years was William O'Brien Donovan. He inspired many with his enthusiasm for the *Grow More Food* campaign, and for the unselfish sharing of his knowledge. This knowledge extended beyond planting — he was instrumental in teaching those who did not know how to make flour out of local produce, how to pickle meat and vegetables to preserve them for future use, and to remember the art of keeping a *pepper-pot* which would preserve meat as long as it was kept going. He also taught people how to make white and yellow butter and a simple cheese known in Venezuela as *Queso de Mano*. The Agricultural Officers throughout Grenada also did their best to ensure that Grenadians produced what they needed. Many of the planters allowed the workers on the estate temporary use of land to grow food for their personal use.

Grenada's extremely fertile soil and abundant rainfall, together with a tradition of small and peasant farming, meant that planting more food for local consumption was no problem. Grenadians already grew a lot of what they ate and they easily increased their capacity to produce. With the new initiative of the *Grow More Food* campaign, Grenadians grew more than they needed, creating a surplus that they exported to nearby islands.

The poorer people in the countryside were accustomed to live off the land and to make only a few purchases from shops. They hardly noticed the disappearance of imported foodstuff. It was the better-off population who now had to eat what Grenada grew and experiment with local food to mimic the missing imported items. The ladies of the house teamed up with their cooks and became experts at making one type of food resemble another. For example, they provided for the elegant table pigeon peas cooked to resemble canned *petit pois*. The cooks of Grenada also used their imagination and ingenuity to use local produce in different ways so as to create a never-before-experienced local cuisine. The delight of the new cuisine eventually lessened the longing among better-off citizens for imported food items.

Shopkeepers, hard hit by import restrictions, managed to survive by retailing local produce, some of which they bought from the community, and some they themselves had grown, for many shopkeepers outside of the towns were at the same time small farmers or dealers for Grenada's export crops. The urban/rural flow of commerce was now reversed, as town dwellers travelled to the rural areas to buy sweet potatoes, yam, dasheen, plantains, and breadfruit.

Grenadian farmers increased the food supply and its variety by planting not only more of what they were accustomed to, but by planting for the first time for a long time some of the traditional crops that had gone out of fashion, but had been staples of the Kalinago and Galibi peoples, the original inhabitants of Grenada. Manioc was once again produced in abundance. There are two types of manioc — sweet manioc and sour manioc. Sweet manioc is white, and could be boiled and eaten like a potato. Sour manioc had to be processed first, to get rid of the poison in the juice. To make it edible, the manioc was grated, and all the juice squeezed out of it, usually through a cotton cloth. The juice was thrown away, and the residue could be used for food. Sometimes the straining was left to "settle", and the sediment used to starch clothes.

Great use was made of the derivatives of the manioc root — cassava and farine. The most common way to use cassava was to make a very filling, flat cake called "Bam-Bam" or "Bambula". This was eaten at nearly every meal including breakfast, and also sold as a roadside food, along with fried breadfruit. A wafer out of cassava was and still is made for ladies' tea tables. Cassava and farine were used in porridge and to thicken stews and gravies. Farine was also used to feed babies. A dumpling called "Pasoo Dumplin" was made out of the sweet manioc.

Judith Parke [24] remembers that, at the beginning of the war, her father used to grow manioc and his mother prepare cassava and farine for sale. People from all over would flock to buy these products. However, the Government soon took over the field and sent inmates from Richmond Hill Prisons to reap the field of cassava. The produce went to feed the prisoners and the inmates of the Mental Hospital. The Government paid her grandmother a fair rent, and her erstwhile customers had to go elsewhere for their supplies.

---

24   Judith, Lady Palmer, wife of Sir Reginald Palmer

Corn, another staple of the first Grenadians, was planted in more abundance during the war, and was used in a variety of ways. It was eaten roasted on the cob, ground into cornmeal that was made into porridge or the more solid cou-cou eaten at the main meal. Green corn was also pounded into a ball, and fried for a breakfast treat. Corn in its many forms as well as peas, rich in protein, took the place of meat at many a meal. Corn flour and cassava were much in demand as a substitute for wheat flour, and in turn became hard to get. Arrowroot, a third food crop of the first people, was no longer produced in Grenada, but ample supplies were brought in from St. Vincent by local boats.

In the rural areas, nearly everyone had a piece of land on which they had tree crops, planted ground provisions, and reared chickens and small stock. Carlyle John spent his childhood during the war in Snell Hall, St. Patrick's. His father's garden provided the family with breadfruit, bluggoe and yams. There were also lots of fruits, and chickens and pigs were reared. In the parish capitals other than St. George's, food was fetched from the nearby countryside. Hermione Greasley, a small child during the war, lived in Gouyave. She remembers that her brother and uncle periodically walked to lands at Florida and came back laden with breadfruit and other staples for the family. Sir Paul Scoon, also from Gouyave, remembers going out to the countryside at 5 a.m. to pick mangoes. He remembers that fish cakes were a special treat reserved for Sunday mornings. He had the responsibility of feeding the family's goat and pig. In his home the usual fare was supplemented at Christmas time with special imported foodstuffs, including tins of corned beef.

Country folk were generous. Families living in the country were accustomed to send provisions by bus into town for their relatives. Now the amounts were increased, and provisions were also sent for persons with whom school children boarded. Lynda Lalbeharrysingh [25] and her sister Cynthia from Sauteurs were pupils at the Church of England High School for Girls. During the week they boarded in town with Mr. and Mrs. Modeste, and would go home every weekend. When they returned, their father, Dolphus Gardner Lalbeharrysingh, would send with them enough provisions and vegetables to last the whole week not just for his daughters, but for the entire Modeste household. Farmers from the

---

25   Mrs. Japal

country, or those having surplus produce from their "gardens" [26], would send food into St. George's for sale. St. George's people who had no one to supply them personally, would go to meet the buses to be the first customers for those bringing plantains, breadfruit, bluggoe, yams, other provisions and vegetables for sale.

Grenada's second island of Carriacou lies twelve miles to the north and is dryer and less fertile than Grenada. Foodstuff, except corn and peas, cannot be produced there in abundance. Every week two or three sloops from Grenada would bring provisions to Carriacou. The entire cargo was sold immediately, and often there were insufficient supplies to meet the demand.

Breadfruit, previously scorned as "slave food" and used only for pigs, was now perhaps the most highly sought-after of the local provisions. Anastasia La Guerre remembers that before breadfruit was accepted as worthy food, it was so cheap that two would be sold for a farthing. The price of breadfruit rose with its popularity. Now two would be sold for one penny. People previously used to put the breadfruit in the bottom of their market basket, cover them with other provisions and vegetables, and "hide to eat breadfruit because only the poor ate it. But now 'Mr. and Mrs. Breadfruit' came into their own, and were carried on the top of the market basket." Besides being cooked as starch for a meal, this fruit came to be used in an almost endless variety of ways. Breadfruit became a staple of everyone's diet and was roasted, fried, baked, boiled, mashed, or chunked for salads. Lincoln (better known as Jack) Baptiste remembered "breadfruit sandwich", made from thin slices of breadfruit instead of sliced bread, with the filling of one's choice in the middle.

That breadfruit could produce a substitute for flour became breadfruit's most cherished attribute. Breadfruit flour was made by peeling and slicing the uncooked breadfruit and leaving the slices to dry in the sun. Having returned safe and sound from Guadeloupe, Eric Glean, an erstwhile inventor, devised a machine that used solar energy to dry the breadfruit slices quickly. But there was no way to mass produce these machines, so only a fortunate few were the beneficiaries of his invention. However they were dried, the slices of breadfruit were pounded to make flour. Breadfruit flour made a heavy, coarse brown bread. Sometimes

---

26   In the Grenadian idiom these are of agricultural land, often at some distance to the house with its own kitchen garden.

the breadfruit flour was mixed with wheat flour if that was available, to produce a lighter, finer-textured loaf. Arnold Cruickshank remembers eating brown bread made from breadfruit flour, and that it was "very tasty", especially when buttered. Arthur Pilgrim remembers that his mother made delicious cakes entirely from breadfruit flour. Alister McIntyre remembers that his mother, Eileen, used to make breadfruit flour, but that she also purchased supplies from the "milk lady" who delivered their milk and who took to making and selling this as a sideline. Flour substitute was also made from plantains, bluggoe and rock figs in much the same way as from the breadfruit, but these flours were not as versatile, and were usually used to make bakes.

Josephine Davis remembers that sugar, although produced locally, was hard to get, and that people in the country resorted to squeezing cane to get the juice to sweeten things. People also used "wet" sugar. Arnold Cruickshank remembers that his grandmother used to buy "wet" sugar for her shop in St. Paul's from an old Barbadian man called Spencer, who had a small mill for grinding cane in that area. Geraldine Sobers[27] recalls that if people still had sweets, they were sometimes used to sweeten tea and other drinks when there was no sugar.

It is undeniable that Grenadians were comparatively well fed during the war, and enjoyed a variety of healthy, nourishing foods. Only a very few continued to mourn the lack of tinned condensed milk. Fresh cow's milk was readily available, even in St. George's, as milk cows were pastured just outside of the town. Butter was made by creaming off the cow's milk, and shaking or churning this with salt and other ingredients. It was customary before the war that local cattle and small stock — sheep, goats and pigs — were butchered and available for purchase by the public on Fridays and Saturdays at various locations in Grenada. This continued, as the advent of frozen meat for Grenadian tables did not occur until the 1950s. Chickens were reared and were fairly readily available for sale, if one did not have one's own fowls running around one's yard. Grenadians always had fish, even if they had to resort to using the lowly Jacks Fish on occasion. Only when purses could not meet the demands on them did Grenadians have to eat meatless meals. Coconuts were also important items of food, and yeast was made from cocoa juice or cassava.

---

27   Mrs. Benjamin

There were, however, some items that Grenadians could not produce, and which remained in very short supply. These included salt, salt fish, rice, cooking oil, matches, potatoes, and cheese. Salt is essential not only for the taste of the food, but a dietary essential for health in the tropics. It was nevertheless only available sometimes, and then was only sold at a half a pound a person. To counter the lack of salt, food was sometimes cooked in sea water. Those lucky enough to have a shop, or know people who did, could use the salt at the bottom of the barrels of pig tails and salt beef. There was a small amount of salt produced at Johnny Branch's salt pond at True Blue, and this was a welcome addition to what was otherwise available.

Fires were kept burning to counter the lack of matches. If your fire went out and you had no matches, a child was sent to a neighbour to "borrow some fire" which he or she would nurse in a pan until it could be used to start the household fire anew.

Some rice was available from Guyana but not on a regular basis, especially after the advent of the U-boats in the Caribbean. Biscuits made in Trinidad were brought over by schooners and were highly sought after, because they were so tasty. They came in large tins, which were equally sought after to be put to all sorts of uses. The rationing of kerosene was particularly hard on the average Grenadian, because this was what was used for cooking and for lamps in the days before total electrification of the island.

The parish of St. Andrew has always been called "the breadbasket" of Grenada, and during the war schooners and sloops would take surplus foodstuff produced in this and other parishes to Trinidad and to other islands for sale. In the rush to seek employment in Trinidad during the war, the Grenville boats carried people as well as produce to seek employment in the big island to the south. Boats also set out with surplus food from the other smaller harbours in the rural parishes laden with Grenadian food to help feed the people of neighbouring islands.

Clothing

From the outset of the war, clothing, cloth, and shoes were regarded as low priority items and were very infrequently imported. These were soon rarities in the shops and more affluent Grenadians learned from the very poor how to "make do". Carlyle John was one of eight children. He remembers his clothes shrank from a good supply to "one set wearing and one on the clothes line". Clothes were handed down through the family, and swapped with other families. When cloth could be had, mothers, seamstresses and tailors sewed for the entire family including shirts for the men. The women also bleached the cotton bags that certain types of food, such as flour, came in. These were used to make sheets, pillowcases, nightclothes and undergarments, especially for the children. If there was time, beautiful drawn thread work, called *hadanga*, was done as a decoration to beautify these items. Joan Parris remembers children of her age on the Grand Anse Estate, which was managed by her father Humphrey Parris, wearing such clothes made out of flour bags.

Providing shoes was a more difficult problem than providing clothes. Judith Parke tells of a shoe sold by Bata Shoe shop during the war that had cardboard soles. These wore out in no time. When shoes began to wear out, mothers put cardboard in the shoes to provide a cover for the holes and protect the children's feet and socks from damage. The services of shoemakers were also very much more in demand. Outgrown shoes were passed to younger children in the family or the children of relatives and friends. To save wear on the shoes, children were allowed to go barefoot more often and on more occasions than had previously been regarded as "respectable".

In a short while, there were no tennis shoes or *watchekongs*, which were canvas footwear similar to tennis shoes, to be had. Youthful tennis players, cricketers and footballers played without the benefit of protection for their feet. Carlyle John remembers that he only had one pair of shoes. In the country, children went to school and everywhere else barefoot as per usual. For example, Anastasia La Guerre, who lived as a child in La Fortune, St. Patrick's, remembers that "she did not grow up with shoes". Some lucky families got parcels of clothes and shoes from their relatives in the United States and Canada, and these were worn by a succession of children until they could absolutely be worn no more.

The shoemakers of Grenada provided invaluable service in mending, resoling and squeezing more and more life out of old shoes. Shoemakers also made new shoes. Norris Mitchell went off to Grenada Boys Secondary School (GBSS) for the first time sporting a pair of shoes made by his father, Cylinford Augustus Mitchell. Cylinford Mitchell was a generous and understanding man, and would now especially repair people's shoes for nothing if they really were needy. Leather was imported from Trinidad, and this was "stretched" with the use of old motor car tyres to make or replace shoe soles. Joe Bailey in Grenville and a Mr. Walker in St. Paul's were two other well-remembered shoemakers of the era. Jenny Yearwood [28] remembers that Mr. Walker made very "pretty little shoes".

Agriculturists who needed to protect their feet against damage in the field were also dispossessed. George Barker-Hahlo remembers that even up to 1948, water boots worn by farmers and field workers were unavailable, as rubber was still scarce. Even if there had been rubber, no shoemaker could manufacture these. Farmers and field workers fell back on the use of the traditional Grenada clog, which consisted of a wooden sole, kept on the feet by a piece of motor-car tyre nailed to the sole to pass across the instep.

Other items were manufactured in Grenada from scrap metal. Chief of these were tin cups, known as pan cups. These were made out of discarded pans and tins. Pan cups were not a new thing to Grenada, but whereas they were once only used by the very poor, in the war years everyone drank out of them due to the lack of crockery. Mr. Marrast was a well-known craftsman who made and displayed his pan cups, cake tins, graters and other items made from discarded tin in his little shop on Market Hill, St. George's. Another famous tin-smith, Mr. Lionel Ferguson, was also known as "Fergie Patat", "Mr. Fergie" or "Papa".[29] He lived in Gouyave, but was itinerant as he would walk all around the island fixing whatever needed to be fixed, so that people did not have to bring their "poorsies" and "tensils" [30] to him. "Mr. Fergie" also made a large variety of items including pan cups, stoves, tin ovens, cake pans, bread tins, funnels, graters and other items from tinning sheets and the large biscuit tins prevalent during the war. He would also solder pots, pans, tin plates and other items that had acquired a leak or hole due

---

28   Mrs. Colin Campbell

29   Lionel Ferguson was also the longest-serving sexton of the Anglican Church in Gouyave. In the days before electricity, he was the town's lamp-lighter. Born in 1906, he died in 2010, close to his 104th birthday. In his maturity he was awarded the British Empire Medal for his services to Grenada. The information on Mr. Ferguson was drawn from the eulogy given at his funeral on 29th September, 2010 by Arthur George Hosten and reproduced in the edition of the Voice Newspaper for 9th October, 2010.

30   The tin household utensils that the poor used.

to long use. Mr. Fergie's journey around Grenada on foot would take several days, and he would sleep at the homes of various friends along the way. Another skilled tin-smith was Mr. Anson in La Borie, who also is remembered as someone to whom people looked to satisfy the need for certain kitchen implements.

School Supplies

Many of the children of those plantation families who were affluent were sent to England to school, sometimes to boarding school, and sometimes to live with relatives and friends. Children were sent away to England to school as early as age six. Due to the elements determining the social stratification of Grenada at this time, these children were almost certainly the fair-skinned "near white". It was generally hoped that children educated in England would remain to make their life there, never, however, forgetting their West Indian roots. Families of this class who could not afford to send their children to England sometimes opted for schooling them in Trinidad or Barbados, which had larger "white" creole populations.

At the outbreak of the war, Grenada boasted of having four excellent high schools, which served the children of the "brown" middle class. The GBSS was exclusively for boys. For the education of girls there was the St. Joseph's Convent, The Church of England High School for Girls, and the Model School, sometimes referred to as Miss Nunez' School. The Model School, with Miss Emeline Nunez as Principal, was classed as a girls' secondary school by the Grenada Handbook (1946). This school included shorthand and typing in the curriculum, in addition to a range of academic subjects.

The standard of education at Grenada's High Schools was so much higher than in the other Windward Islands that many wealthier coloured families from St. Vincent, St. Lucia and Dominica sent their children to school in Grenada, either as boarders at the St. Joseph's Convent or GBSS, or to live with family and friends. There were also 54 primary schools in Grenada, five being located in Carriacou and Petit Martinique.

Secondary school supplies and stationery were in short supply during the war years. Janice Bain and Marcella Lashley remember that erasers and exercise books were hard to get, as were materials for art and craft. Textbooks had to be bought from students in the year ahead, and at the beginning of the year there was always "a bit of a scramble" by parents to make sure that a schoolchild was satisfactorily provided with the necessaries. Sometimes school books and supplies were brought for pupils in Grenada by persons who travelled on business to Trinidad or further afield. Parents also sent "messages" [31] regarding school books and supplies with schooner captains who plied the Grenada-Trinidad route.

The uniforms of the Church of England High School for Girls and the St. Joseph's Convent originally called for the wearing of "Panama" straw hats which could be re-blocked and refurbished as necessary. Bertie Andrews was one of the people who re-blocked school hats. Hats were set on a frame and painted with sulphur and other chemicals to make them both keep their shape and be as close to white as possible. These imported hats became unavailable as soon as the war started, and hats of local straw were substituted. However, these could not stand up the daily wear, tear and tussle of the schoolgirls, and rather than let girls appear in shabby hats, the headmistress of the two girls' schools decreed that hats should only be worn on special occasions. Hats were soon after dropped entirely from the school uniform ensemble. Therefore, a minor repercussion of the war was that the pupils of the girls' high schools ceased to wear hats as a regular part of their uniform. Unfortunately in those days the link between too much sun, skin cancer and cataracts was unknown.

Somehow, however, the mothers of the nation managed to keep both boys and girls properly dressed in the traditional school uniforms throughout the war. A girl had a solitary serge skirt that would be cleaned during the school holidays. New skirts were made long with big hems, so the girls could "grow into them". Outgrown skirts and blouses were also passed from one child to another within family and friendship circles. The same was true for the boys, except that boys, being boys, needed a spare pair of pants. As boys' clothes did not last as long, there was less to pass down.

---

31 In Grenada a shopping request is commonly called a "message" — hence "to make a message" means to be sent to buy something.

Motor Vehicles, Spare Parts, Tyres and Gasoline

New motor vehicles were basically unavailable in Grenada once the stocks ran out, and they were not again easily obtainable until several years after the end of the war. Those vehicles that were available at the beginning of the war and vehicles that people desired to import had to be applied for, and even then permission for sale or import was only granted after "fullest justification and details of use" were provided. In a news item on this subject in *The West Indian* of 19th August 1944 it was stated that "no undertaking can be given that requirement will be fully met" to those applying to import vehicles.

Tyres for motor vehicles and bicycles were, after a while, also not available. A notice captioned "Recapping of Tyres" appeared in *The West Indian* newspaper of 11th November, 1944. It stated that:

> With the approval of the Transport Emergency Board, Trinidad, arrangements have been made with Messrs. Charles McEnearney & Co. Ltd. to recap a limited number of tyres during the next 12 months. For the present the Competent Authority will collect tyres to be recapped, ship them to Trinidad and when they are returned hand them back to the owners after payment of all expenses.
>
> Only tyres used on essential vehicles will be accepted for recapping.

The scarcity of tyres and inner tubes meant that vehicle owners had to use their tyres and tubes far beyond what was safe. Some of the tyres were so smooth that they constantly got punctures. A trip by bus was interrupted several times by flat tyres that had to be patched before the journey could continue. A trip from St. Andrew's to St. George's could take as long as eight hours. When the tubes absolutely could not be used any more, they were removed, and the tyres stuffed with straw. Bolts were passed through the tyres to keep the stuffing in place. Needless to say these tyres gave a very bumpy ride!

Apart from the lack of tyres and spare parts, gasoline was strictly rationed, and a permit was necessary for both buying and selling. Gasoline was only provided for the essential services, and for individuals who absolutely needed to have their own means of transportation. As shortages increased, there were fewer and fewer vehicles on the road, because one by one the vehicles were garaged for the rest of the duration of the war. Even the bus service, which was used extensively by the travelling public, was curtailed on some routes due to the shortages of gasoline, spare parts and tyres. One route that ceased operation was that from Birchgrove to St. George's. After no more busses plied this route, people simply had to walk.

Josephine Davis remembers that one of the only buses running in St. David's during the war was the *Excelsior* owned by Mr. Joseph Gibbs of Crochu. The van carrying mail to the rural parts also carried passengers, but seats on these had to be booked two weeks in advance. A regular bus service from Sauteurs to St. George's was operated by John Watts' father, Cecil Sitwell Watts, using two buses: *Civility* and *Grand Central*. There was no need to book a seat, but passengers had to be on the roadway in Sauteurs by 7 a.m. The bus made scheduled stops until it reached St. George's. In the afternoon it would begin the return journey leaving at 2 p.m. Watts also owned a gas station in Sauteurs which supplied both gasoline and kerosene when it was available. Both gasoline and kerosene were brought into Grenada from Trinidad by schooners and sloops. For any reasonable distance, Grenadians returned to the abandoned habit of getting where they wanted to go by walking. Jenny Yearwood remembers that due to the infrequent bus service, she walked to The Church of England High School for Girls from her home in Mardi Gras, leaving home at 8 a.m. to reach school a few minutes before 9 a.m. She walked back in the blistering three o'clock afternoon sun. At this time, The Church of England High School for Girls was housed in Lamolie House [32], located on the corner of Church Street and Young Street, where the main branch of First Caribbean International Bank now stands. The walk took forty-five minutes and was made less tedious by the company of her sister Jessica and cousins, who also daily walked from home to school and back again.

Luxury Items

If clothes and shoes were regarded as low priority items for import during the war, corn flakes, chocolates and manufactured sweets were regarded as absolute luxury items and were not on the import list. They were in evidence only when visitors or travellers brought them in as part of their luggage, or they came in parcels sent by relatives abroad.

---

32   Grenadians recall with glee the story that this building was haunted by the ghost of Mr. Passey, who was frequently seen by students. This ghost is supposed to still haunt the bank building, which was built on the foundations of Lamolie House.

Arthur Pilgrim, a boy during this time, remembers being given a box of chocolates as a gift for his holiday in St. Vincent. The temptation of having these beloved and scarce items in his hand was too great, and he "pigged out" on them, finishing them during the voyage on the *Marcelle S*. As a result he soon began to feel very, very queasy, although he claims that he did not "feed the fishes" at any time during this journey.

Certain toiletries also disappeared from the shelves. "Merry" Meredith McIntyre, a chemist and pharmacist, obtained from his friends and colleagues in Grenada and elsewhere the formulas for tooth powder, talcum, soap and bay rum. These he produced during the war years as acceptable substitutes for the imported versions no longer available.

In the years before the war, children in Grenada did not have the plethora of toys that they have today. Most toys were made by the adults for their small ones, and the bigger children made their own playthings. During the war years, home-made toys again became the only ones available as during the war and the years immediately after, toys were not imported. Some of the cuddly toys made by Grenada's womenfolk during the war were truly magnificent. Evelyn Pilgrim made pandas and teddy bears from special material procured through the efforts of her husband, Mr. H.H. Pilgrim, the Inspector of Schools. These bears were jointed, the joints carefully crafted by her son, Arthur. Mrs. Pilgrim also made dolls. The bodies of the dolls were of cloth, and the faces were made of *papier-mâché* which was pressed onto a mould, dried and painted. The bears and dolls were sold at Granby Stores and R.M.D. Charles at Christmas. Carmen Alexis of Morne Jaloux was also an expert doll-maker. Beatrice Bain also made beautiful dolls for her children. Some mothers were quite capable of painting on the doll's face, but sometimes an "expert" was asked to do this delicate job. Kathleen Bain was one of the lady experts for the painting of dolls' faces. Older girls made their own dolls out of corn cobs with the corn 'beard" for hair. Dolls would be dressed in clothes made from scraps of cloth. Brenda Wells [33] won her baby doll called "Mary" in a raffle at a The Church of England bazaar when she was eleven. This was the first imported toy she ever had. Boys would play with box carts with cotton reels or ball bearings for wheels. They also made scooters out of scrap wood, with wooden wheels or ball bearings to make them go. Another favourite homemade toy for boys was the wooden top.

---

33   Mrs. George Williams

Imported toys were missed particularly at Christmases during the war, because children were always accustomed to get a manufactured toy or two only during this season if the family could possibly afford this. If a child got an imported toy during the war years, this was usually at Christmas, and sent by family living overseas.

In order to have toys to display and to sell for the festive season during the war, stores in town employed young men to make toys which the stores could then offer for sale. This was the first time that "ready-made" toys displayed in the shop windows were manufactured in Grenada! The Hood boys were among those commissioned to make wooden toys for R.M.D. Charles Store for sale at Christmas. Beautiful wooden toys were also made by the Misses McLeod of Tufton Hall Estate and were sold at the Home Industries.

A number of other items taken for granted in peacetime were also regarded as non-essentials and completely disappeared from shops and stores for the duration of the war. When Eileen McIntyre and Alan Gentle got married during the war on the eve of his departure to take up his commission with the Windward Island Battalion in St. Lucia, Eileen and her family had to be inventive in ensuring that her wedding did not lack the essentials. She wore her mother-in-law's wedding veil of Brussels lace. Eileen's godmother made the almond paste for the cake out of local almonds. Hundreds of almonds had to collected, the nut removed and the soft seed peeled to be crushed and made into paste. This was truly a labour of love! There was no wine to be had, but the day was saved by a gift from a priest of a case of wine! Jewellery was another luxury item which was not imported during the war years, so wedding rings could not be purchased. Eileen tells how the wedding rings for her and Alan were made.

> My wedding ring was made by a local jeweller from a small bar of gold which my mother-in-law kindly provided. He also made a ring from two half sovereigns my godparents had given me at birth, and this I gave to Alan. [34]

Without the ability to source jewellery and silverware from abroad, the skill of local jewellers was much more appreciated. One of these local jewellers was Mr. John Fleming, who made trophy cups, plaques, rings and other items out of precious metals. Other Grenadians became experts

---

34 Gentle, P. 70

in making costume jewellery such as pins, necklaces, earrings, bangles and other items for the ladies out of various materials. Jack Baptiste began making initial brooches for the ladies out of wood, and painting them as a hobby during his recuperation from rheumatic fever. The brooches proved so successful that he developed his hobby into a small industry, expanding into the production of brooches, earrings and other items out of turtle shell. Jack's turtle shell jewellery remained popular even when manufactured jewellery was once more available after the end of the war, and his handmade items became prized possessions.

VIII: CONTRIBUTING TO THE WAR EFFORT

War was enormously expensive to Britain; it also meant additional expenditure for Grenada. The Censorship Service, the Competent Authority, relief to Government employees on military duty, a First Aid and Ambulance Detachment, medical emergency supplies and black-out material for Government institutions — these were all costs that Grenada had to meet locally.

This extra expenditure was met with an additional export tax on nutmegs, mace and cocoa. Government revenue had fallen due to the restrictions of imports and the resulting loss of import tax. For the first years of the war, this extra tax on exports could be met without too much hardship. The economy was not too badly off, as the price for nutmegs and cocoa, which had slumped in 1938, rose again during the war. These exports could thus stand the extra tax, which only affected the planters and the wealthier sectors of the Grenadian community. This was explained carefully to T.A. Marryshow, whose concern was that the cost of living "of the man lower down" [35] must not be affected.

Not only was Grenada asked to raise Government income to meet the extra expenditure in Grenada due to the war, but the society was also called upon to raise money for the war itself. A War Purposes Committee was set up in Grenada to collect funds raised in the society and to decide on how these funds would be spent. A total of £20,069 18s and 3d

---

35 Popham to McDonald CO 321/386 1939.

was collected. This amount was used for various war purposes, with sizable donations to the Red Cross and other charities. The "big ticket" items purchased with the fund from Grenada was one fighter plane, and half of the cost of a mobile canteen. Donations were also made to the West Indian War Services Fund and to the British Red Cross to support charities that helped people injured in the war, and to pay for the training expenses and overseas travel for Grenadians recruited for the Royal Air Force (RAF). Funds were raised for this not only by quite large donations but by all sorts of small activities by Grenadian people of "all ages and stages".

Grenadian women were particularly active in raising small amounts of money for the "War Effort". As a girl, Oris Teka [36] and her friends sold flowers every other Saturday morning. The flowers were donated by those who had gardens, and the girls would make them into bunches, and place them in a basket with a red bow. Oris remembers that the bunches were readily sold, and that donations were also given to the girls by persons who did not take the flowers but left them to be sold once more to someone else, so that a little more money could be raised. A morning's work netted the girls less than one pound sterling. This money was given to Laurie Stephenson at *The West Indian* newspaper, who passed the money on to the appropriate authority. Bake sales, raffles, and sales of work were also held by members of the public to raise money. The sales of work featured the beautiful embroidery for which Grenada used to be famous, other hand work and bed linen made from bleached "flour bags" decorated with hadanga. *The West Indian* newspaper would from time to time publish the names and amounts of contributors to the "Invasion Fund".

As well as contributing to the sales of work, middle class housewives knitted blankets, balaclavas, scarves, gloves, mittens and socks for the troops overseas, and they taught their daughters and sometimes their household help to do the same. Mrs. Ormiston, the wife of Major Ormiston, the English Chief of Police, organised a group of girls to knit woollen squares that would later be joined together to make blankets for the forces. Some boys were also pressed into service at home by their mothers to knit these squares for blankets.

---

36   Oris Lady Graham, wife of Sir Samuel Graham

Many of the women had witnessed their own mothers engaged in activity during the First World War, and had learnt the craft from them. Now it was their turn. Julian Rapier remembers his mother, Hilda Rapier, knitting and making bedspreads — all to be sent to England. Along with Ethel Mary Donovan, Elma Comissiong, Jane Mahy and several other women she was presented with a medal from the Red Cross at the end of the war. This was in appreciation of the quiet contribution of these women in providing warm clothing for the troops on the European front. Supervised by teachers, schoolchildren from both secondary and primary schools also knitted socks, vests and caps for the troops. Students of St. Joseph's Convent remember the nuns knitting while they supervised their pupils in this exercise. A former pupil of the St. Patrick's Roman Catholic School remembers the girls set to knit items in brown wool and white wool as part of their school activities.

Scrap iron was collected by the Government and sent to England for the war effort. Many of Grenada's beautiful water wheels were dismantled and destroyed for this purpose. Randolph Mark in his pamphlet on the Mount Carmel Waterfall states that the most elevated part of Post Royal Estate was fortified during the 17th and 18th centuries. The cannons from these fortifications were removed during the Second World War to be used as scrap iron. [37] Godwin Brathwaite remembers that some of the machinery used in Carriacou to manufacture lime juice was collected and shipped to England as scrap metal. Carlyle John remembers that, as a pupil of the St. Patrick's Roman Catholic School, the children were asked to collect scrap iron wherever they found it, and to bring it to the school where it would be collected by trucks and shipped away. Cecil Edwards remembers collecting crown corks and old metal to assist in this effort. Harry Ogilvie adds that some children collected silver paper from cigarette packets and rolled this into balls. This they donated to the war effort. The children of Grenada were told by their teachers that the scrap metal was used mainly for the manufacture of ammunition.

Another way of raising money to finance the war was the issuing of War Bonds, in which many Grenadians, who could afford it, invested. School children also donated and collected money for the war effort. Leo Cromwell remembers putting a percentage of his pocket money every week in the school fund for the war. Young people also got together

---

37 Mark, P. 9

to form flute and other types of bands, and went parading through the streets of the parish capitals, entertaining, and collecting money. Bags were attached to long poles so that money could be collected from onlookers on first floor balconies easily as the parade went by. Godwin Brathwaite and Robby Rowley, who attended the Hillsborough Government School, remember going around collecting money in Carriacou "to buy a *Spitfire*". Nearly everyone was engaged in some way in helping to fund World War II, and for such a small and relatively poor island, Grenada and its people provided Britain with substantial monetary assistance for the war effort. Grenada in total collected a sum equal to that of Malta.

IX: GRENADIANS IN MILITARY SERVICE

Early in 1939, in anticipation of war, the Grenada Volunteer Force had been amalgamated with the Police Force, renamed the Grenada Volunteer Force and Reserve, and put under one Commanding Officer. This is the force that was called into service on the first day of the war. The Cadet Corps was a part of the Volunteer Reserve, and made up mostly of younger men. Their uniforms were of khaki, and there was a shooting range at Queen's Park where the Cadet Corps practiced with old Lee-Enfield 303 rifles. The practices were not held very often, due to the shortage of ammunition. Dennis Malins-Smith was a cadet, and remembers practicing shooting up to 600 yards. A brilliant musician, Dennis also played the trumpet in the Cadet Corps Band. Otherwise, Dennis worked at *Everybody's Store* in charge of the Hardware Department.

The Volunteer Reserve was trained by Captain the Hon. Earle Hughes who had been an officer in World War I. The men in the Grenada Volunteer Force and Reserve were given some military training and equipped with uniforms and rifles. Members of the Volunteer Reserve were assigned to patrol the streets of St. George's by night, walking in pairs.

However, the first reaction of many able-bodied men in Grenada to the declaration of war was not to clamour to join this force, which was tantamount to a Home Guard, with the responsibility to protect Grenada and Grenadians on the home front. Rather, they looked for ways to go to the Mother Country with the objective of fighting the enemy in Europe.

This enthusiasm to go to England to fight was somewhat cooled by the statement from the Colonial Office that the best contribution Grenadians (and people of the other colonies) could make was to stay at home and keep on producing the raw materials and food that Britain needed. The Colonial Office advised that Grenadians should join the Grenada Volunteer Force and Reserve and serve by protecting Grenadians at home from any enemy action, for no part of the British Empire would be free from war, and they could also serve by providing protection for their own territories. In spite of this statement, twenty-one young Grenadians, mostly but not exclusively of the middle and upper classes, paid their own passages to England or Canada to join the armed forces of those countries. Some of these young men came from military families where their fathers had served in particular units of the British or Canadian Armed Services. But the removal of these twenty-one educated Grenadians left the Grenada Volunteer Force and Reserve short of educated men to recruit as officers, as there was not a large pool of educated men of the right age from which to draw. [38]

Some of the well-known Grenadians who served in the British or Canadian Armed Forces overseas were Douglas Gordon Alexander, Michael Bain, Leo deGale, Terry Evans, George Graves, Gordon Haydock [39], Keith Mancini, D E.W. Rapier, and Vivian Williams. Five women, Myra Woodruffe, Rita Kerr, Margaret Gun-Munro, Leah Bascus and Betty Kent [40], were accepted by the Auxiliary Territorial Service (ATS) in England. Olga de Gale was a civilian firewatcher in London. Firewatchers performed a vital service during the war in England, in view of the number of incendiary bombs dropped by the Luftwaffe. People designated as civilian firewatchers were required to report for duty on a regular roster. Olga deGale was killed by a flying "doodlebug" bomb in Regents Park, London.

---

38   Popham to McDonald CO 321 386. 1939
39   From Carriacou
40   Mrs. Mascoll

The Nursing Service in England was also augmented with several Grenadian nurses, including Monica Monroe [41]. Eight Grenadians were among the West Indians who worked in Britain as munitions workers. Many others also paid their own passages to England, expecting that the war would provide expanded employment opportunities there, superior to what prevailed in Grenada.

Apart from the Grenada Volunteer Force and Reserve, Grenadians could serve in the Trinidad Volunteer Naval Reserve. This was formed in 1939, and recruited its membership from all over the Caribbean. Thirteen Grenadian seamen were included in the ranks of this Reserve.

In October 1942 all local forces were taken under Imperial control with headquarters in Jamaica. Under the control of headquarters were the Northern Caribbean Force, also headquartered in Jamaica, comprising battalions in Jamaica, Bermuda, Bahamas and the Leeward Islands, and the Southern Caribbean Force, headquartered in Trinidad, with battalions in Trinidad, Barbados, and the Windward islands. The Northern and Southern Caribbean Force were under the command of both English and Caribbean Officers. Grenadians enlisted as both officers and ordinary ranks. An arm of the infantry and an arm of the artillery of the Southern Caribbean Force were stationed in Tanteen, St. George's, where a camp for 44 infantry was built on the site that had been prepared for the Grenada Boys' School. The Southern Caribbean Force was poorly equipped, mainly with what was left over from the other theatres of war. It functioned mainly to provide psychological comfort to the population, because in no way could it defend against a well-equipped invasion force.

Coastal batteries were stationed at strategic points around St. George's, especially at points where there were vulnerable targets. The Seventh Grenada Coastal Battery was stationed on Ross's Point. Herbert Payne, a Trinidadian of Grenadian ancestry, was in command of this battery. Cecil Harris from Mount Moritz is a veteran who remembers this firm but gentle giant of a man. Cecil Harris, Clive Medford, Mottley Hinds, Cogland Searles and Arthur Searles were all friends from Mount Moritz, St. George's. They all enlisted together in the Southern Caribbean Force, and all made it through the selection process and medical. After the training, which they all enjoyed, these five friends were assigned as

---

41   Mrs. George Clyne

gunners to man the "25 pounder" emplaced on the precipice at Ross's Point. Whenever there was a drill or a real alert, everyone had to rally as soon as possible. Cecil Harris recalls that they often went to bed in full uniform in anticipation of being called up by First Lieutenant Payne in the middle of the night. When they were on "Alert A", everyone had to appear in full dress with their helmet on and carrying their Lee-Enfield rifles when the muster was sounded. At other times, they could appear how they were. Sometimes Mr. Payne sounded the muster five times in one night, and at other times left them to sleep in peace for several nights in a row. This was to ensure that they were kept alert at all times.

In certain types of drills, the command was given to man the gun. There were five functions for the gunners, and each man learned every function. When an alarm was sounded, whoever reached the gun first filled the most urgent of the tasks. The places were interchangeable, but always one took the sights, one loaded, one took the orders, one was there to fire, and one to take out the empty shell. But the gun was never fired. There were also daily exercise sessions and drills. Cecil and his friends thought that these "were fun".

The soldiers at Ross's Point were well fed, but the meals were not very tasty. Cecil remembers that once they were served rice that was full of weevils, but they did not mind too much, for at least they had rice when the general population could buy none at all. Every Christmas the soldiers were given a really festive dinner which they thoroughly enjoyed.

The soldiers were given "hard tack" biscuits when they went on marches to various parts of Grenada. They particularly enjoyed a march to Gouyave, as the soldiers were greeted so joyfully by the population, who were glad to see those protecting them with their own eyes. The people of Gouyave brought out many water coconuts, which were a welcome treat for the soldiers. Cecil Harris also served in Trinidad, and remembers marching on parade down Frederick Street.

Among the other Grenadians who served in the Southern Caribbean Force as officers or rank include Ronald Wells Bain, Claude Bartholomew, Cosmos Cape, Henry Christopher, Allan Gentle, Gordon Haydock, Roy Hughes, Ben Jones, Derek Knight, Dennis Lambert and Rooney Mauricette. Grenadians in the Southern Caribbean Force served not only in Grenada, but also in several islands in the Caribbean.

Several women from Grenada, including Cynthia Salhab, Eileen "Chickie" Moore, Elaine Moore, Albie Fletcher, Muriel de Riggs, Ruby Shillingford and Dora La Grenade [42], served in the Caribbean Branch of the ATS, and were employed mainly as secretaries performing clerical duties. However, the ATS also did surveillance duty on foot and by van, especially in the country where they had to cover long distances. Groups made up of young women from the different Windward Islands were stationed in Grenada, St. Lucia, St. Vincent and Dominica. They wore a khaki uniform that was a tropical version of uniform of the British-based ATS. They were said to look "very nice". Their service is proudly remembered.

Although Britain did not wish to recruit soldiers from the colonies early in the war, this was not the case for air force personnel. As Britain pounded the "Invasion Coast" to disorganise Hitler's military aspirations, the losses of aircraft and crews were great. Over one thousand RAF planes and crews were lost in one battle alone, this while supporting the evacuation of Dunkirk in May 1940. Farfan describes the continual loss of young airforcemen during the war:

> The statistics of survival were at long odds against us. Through 1942 and 1943 and up to March 1944, Bomber command's attrition rate averaged out at 5% per operation. This is what I met. 85% of all crews were lost before completing thirty operations. Survivors were the exception. On the average 65% of the operating aircrew would be killed, 20% would be POWs, and 1% would live through their tour of thirty missions.[43]

The British Empire was called upon to provide crews for the new planes that the USA provided. Caribbean people, along with others from the British Commonwealth, were recruited for the RAF as pilots, flight crew, mechanics and maintenance crew. Applicants from the Eastern Caribbean and Trinidad were screened and processed at the RAF base in Trinidad.

Each applicant went through a rigorous medical examination and educational test before being selected for any part of the service. Selection as a pilot or flight crew was dependent on higher educational qualifications and aptitude. Those who were selected as possible pilot

---

42   Dora la Grenade died in service at age 19.

43   Farfan, P. 218

material went through an initial training scheme at the RAF base at Piarco. Those who made it through the initial training were sent to England for further training. For the first wave of recruits, the initial training was repeated in England until the exigencies of the service and experience of the quality training of pilots trained in Trinidad convinced the RAF that the training in Trinidad was as good as that offered in England. One of the pilots recruited in Trinidad was Julian Marryshow, the son of T.A. Marryshow. Julian writes:

> It had been a closely guarded secret that Trinidad was the first RAF Pilot training scheme outside of the United Kingdom. I applied to join the Air Cadet Trinidad Scheme. The first course had already been selected and had commenced training. I was among dozens of applicants for the second course, and after a rigorous series of medical tests and interviews, I was among the twelve selected. [44]

After being transported to England "on a tanker carrying high octane fuel" and receiving further training, Julian was assigned to fly a *Spitfire* and later a *Typhoon* fighter bomber. He survived without injury when he was shot down over Holland, and returned from the war definitely the dashing hero. Others recruited for the RAF as pilots or crew included Jackson Dunbar "Jack" Arthur, Michael Anthony Cruickshank, Joseph Ferris, William Grahame Lang, Colin Ross, James "Jimmy" Ross, and Harold K. "Buzz" Shannon. Messrs Leslie Seon, Harry Noel, Raeburn and Benjamin and others were recruited to serve in the RAF as aircrew, tradesmen or munitions workers. [45]

By 1942, with Britain standing alone against Germany and her Allies, the pressures on Britain's resources were so great and the casualties so high that the War Office in Britain sent out a call to all Britain's colonies for volunteers to enlist in the British Army. At last, Grenadians got their wish to go abroad to fight for King and Country in foreign lands. In 1944, a Grenada Detachment of the Windward Islands Garrison of the Caribbean Regiment was established for the additional protection of the eastern and southern Caribbean and for possible service overseas. There were 139 recruits for Grenada Detachment of the Windward Islands Garrison. Grenadians of all walks of life and from all over Grenada responded to the new recruitment drive for men to serve overseas. In Birchgrove, 24 young men were recruited. This was the most from any

---

44 Metzgen and Graham, P.148
45 For an account of Grenadians who served, see Brizan's *Brave Young Grenadians.*

one village.

For most people who enlisted there were mixed motives. Undoubtedly there was loyalty to Britain, and fear of what it would mean if the Axis countries were victorious. But, although they wanted to help England win the war, they also wanted to escape from the lack of opportunities in Grenada, and to better themselves through the opportunities that military service would bring. They wanted to broaden their horizons through travel. Enlistment also meant steady pay and a chance to see somewhere else besides Grenada. They were willing to risk their lives to do this.

Selection and training took time. Recruits served in the Caribbean until they were called up to go to Britain. However, they were warned that they could be called to travel at any time. Their local training consisted of five months training in St. Lucia, a further training period in Dominica, and final training in St. Lucia, after which a selection was made among the men, and these were sent to Trinidad for medicals. Only those who passed the medicals were earmarked to be sent overseas.

It was not until June 1944 that two contingents of troops of the Caribbean Regiment arrived in Britain. Another contingent arrived in early 1945, with Grenadians included in each. The first group of 20 Grenadians to serve overseas with the Caribbean Regiment included Privates Alexis, Andrew, Barriteau, Callendar, Charter, Chandler, Cox, Cuffy, David, Edwards, Gibbs, James, Roberts, Louison, Lander, and Sgt. George Phillip. Most were sent to Italy in a mixed Company with other Caribbean soldiers, mostly from Guyana and Jamaica. In Italy, Caribbean soldiers were assigned routine garrison duties as well as escort duty for German soldiers captured on the battlefront in Italy to prisoners-of-war camps in Egypt. Sgt. George Phillip and Pte. Ronald Ivan David were among those assigned to guard and escort duty. David remembers that all the German prisoners-or-war could speak excellent English.

The Caribbean Regiment continued to serve in Italy and Egypt for a short time after the war officially ended. All soldiers were granted leave before they went home, and provision was made for them to use their leave to tour either Egypt or Israel. David chose Israel and enjoyed a guided tour of the Holy Land, mounted especially for troops. Irie Francis, who had enlisted in Trinidad and served in a separate Company, chose a holiday camp in Egypt run for the recreation of servicemen. Before too long, however, the soldiers from Grenada were on their way to St. Lucia where they were disbanded. Although primed up psychologically to fight, Grenadian lives were to be in minimal danger. They did not see the action they expected on the battlefields of Europe. Men were still being recruited and trained for a Second Caribbean Regiment when the war ended. Grenadians who served abroad acquitted themselves well, in spite of some incidents of race prejudice from the British military officers, the British Government and from some of the British public. There were also incidents between the Caribbean troops serving in Italy and white soldiers from South Africa who were also stationed there, who thought that Caribbean soldiers should not be allowed to carry weapons. *The West Indian* newspaper, which kept the Grenadian public abreast of the news of Grenadians serving abroad, printed a news item on 30th May, 1943, which read:

> Corporal D.E.W. Rapier, of the Canadian Corps of Signals, who was heard on the air in BBC's Calling the West Indies programme last Tuesday evening, has recently passed his first class Signal Operators test. In a recent letter to his parents, Mr. and Mrs. Cecil W. Rapier, of Grenville St. Andrew's, he said he was the first member of his division to pass this test.

Only three Grenadians did not return from the War: Sgt. (Air Bomber) Joseph Ferris who was killed on 15th October, 1942, Flying Officer (Air Bomber) Colin Patrick Ross who was killed on 3rd November, 1943, and Rear Gunner Sgt. J.D. "Jack" Arthur, who was killed on 17th September, 1943. These three were all members of the Royal Air Force who lost their lives in separate incidents in the skies of Europe. As sad the loss of these three was, Grenada was fortunate not to lose more of her sons serving in the RAF. Esmond Farfan has this to say about the families of RAF personnel from Trinidad:

> None of those we had left behind expected to see any of us again, and half of us who had left for war never did come back.[46]

The names of the fallen were inscribed on the new War Memorial that stands in the Botanic Gardens and Ministerial Complex in Tanteen. Their graves in British military cemeteries in Britain, France and Germany are scrupulously maintained by the Commonwealth War Graves Commission. On 19th July, 1944, Olga deGale, a civilian firewatcher,

---

46   Farfan, P. 532

was injured when a V1 flying "doodlebug" bomb damaged the part of Hanover Lodge where she lived and worked. She died the same day. Hanover Lodge was repaired and stands today on the Outer Circle of Regents Park, near London. Miss DeGale received a commendation for her work by the Commonwealth War Graves Commission. A few people still remember and respect those who gave up their lives for Grenada, the Empire, and a philosophy of life precious to all Grenadians.

X: SPIES

Early in the War, the Colonial Secretary had warned that no part of the British Empire would be untouched by the war. This was to be so very true of the Caribbean. In their planning for war, Germany recognised the strategic importance of the Caribbean, while the Allies did not. Germany laid the foundation for a successful campaign in this region. When war was anticipated, but not yet begun, some among the population were recruited as German agents to collect sensitive information on ship movements and other intelligence. They would transmit all information deemed useful back to Germany, usually by radio. There were Caribbean people who were German sympathisers who were willing to perform this function, and others who would do it simply for the money they were paid. There were German agents serving on neutral ships gathering intelligence and making contact with local spies. Germany had a well-developed intelligence network operating for years before the declaration of war on 3rd September, 1939. Germany continued to gather information from their informants all during the war. As hostilities became imminent, neutral ships coming into Trinidad were intensely scrutinised by British Intelligence, but the contact between the crews of these vessels and local sympathisers continued unchecked. Germany, through this means, had up-to-date information on the developments in Trinidad.

When war was declared, the most potentially dangerous German sympathisers were interned in a War Detention Centre within the St. James Barracks area in Port-of-Spain, Trinidad. This facility was made very secure, and also housed shipwrecked German officers and seamen washed up on the shores of Trinidad, Grenada and the islands of the

Lesser Antilles. Germans and Italians in the Caribbean and all who acted suspiciously were detained and sent to the St. James facility as soon as this could be arranged. These prisoners-of-war were detained in this facility until the war ended.

Another detention camp was on Caledonia Island, one of "The Five Islands" – a group of six small islands lying west of Port-of-Spain in the Gulf of Paria. Persons deemed less dangerous, including non-military persons from the Axis countries deemed to be a danger to security, were interned here. Known political activists such as Buzz Butler were also incarcerated there for the duration of the war.

In spite of these measures, German agents operated throughout the war without being detected, sending intelligence to Germany. In Trinidad, there were several serious incidents, including Morse code signals being seen transmitted from locations near Manzanilla on Trinidad's east coast on the nights of the 25th and 29th July, 1943. In one instance, the signals were answered purportedly from U415 offshore. These incidents caused "a major flap", with troops sent out to cordon off Manzanilla beach, but the signal senders were never caught. [47]

At the beginning of the war, the Grenadian population was warned by the Police Administrator to be sensitive to the likelihood of spies in Grenada, not to befriend any strangers, and to report the presence of all foreigners to the police. They were also asked not to talk about movement of ships that they observed or about which they had incidental information. The slogans "Careless talk costs lives" and "Loose lips sink ships" were well known in Grenada.

Grenadians, including the children, were sensitive to anyone who seemed to them to be suspicious. As a result some very real spies were caught, but some embarrassment was suffered by innocent people as a result of over-enthusiasm on the part of the population who suspected every white foreigner, known or unknown, of being a spy.

But there really were spies in Grenada. The best-known tale concerns two men claiming to be North American wrestlers, calling themselves Joe "Whiskers" Blake and Joe Gotch, who arrived in Grenada during the war and temporarily settled down here. They staged wrestling matches in The Church of England hall for the entertainment of the public, but

---

47   Kelshall, P. 366

"Whiskers" also spent a lot of time on the beach spear-fishing. He taught many of the local youngsters to spear-fish, but it was discovered that his fishing was a cover for observing ship movements. He never let locals into his house, not even the boys whom he had befriended and taught to fish. His laundry line was discovered to be the aerial for his secret radio transmitter when one of the youngsters received an electric shock when he accidentally touched it. On investigation, it was discovered that the radio transmitter was disguised as a refrigerator. Needless to say, both men suddenly disappeared. Later, when the credentials of Joe "Whiskers" Blake and Joe Gotch were checked, there indeed had been American wrestlers by these names, but they were long deceased! "Whiskers" Blake is alleged to have been arrested later in Trinidad as a spy so important that he was flown out to Great Britain rather than being detained in Trinidad.

There was a foreigner known as Miss Gast who stayed at the *Hotel Antilles*. She got herself invited to a lot of parties, but it was soon noticed that she was too interested in certain types of information. No one knows what happened to Miss Gast, but she mysteriously disappeared from the social scene. Spies were identified in Carriacou as well. Robby Rowley, son of Redvers Rowley, the District Officer for Carriacou, remembers that his father's radio provided news of the war from the BBC, not only to the family but to the community. It was also a source of entertainment, providing classical music for the listening pleasure of the family. One day the radio ceased to work. A day or two later, a white foreigner turned up at the house, saying he was a radio man, and suggested that Mr. Rowley let him look at the radio. Thus he set to work and fixed it easily. Then he disappeared. The question in people's minds was: was he a German spy or simply a yachtsman travelling through the Grenadines?

Grenada and Grenadians owe a belated apology to those innocent people who were thought quite erroneously to be spies. Among those who were caught in the net of over-enthusiasm was Fr. Raymond Devas, O.P., the much beloved English Dominican Priest, a keen naturalist, ornithologist and author of several books on Grenada. During the time he was assigned to the Sauteurs Parish Church, St. Patrick's, he was observed going off on his motorcycle, and seen in lonely places looking through his binoculars. Accused of spying, investigation proved that he was only bird-watching! Undeserved suspicion also fell on a young agricultural student named called Pedrito DaSilva. This popular young man, the son of a wealthy land-owner in Brazil, came to Grenada to study the cultivation of Grenada's special strain of cocoa. He pursued his studies and had a wonderful time in Grenada as well. When he was urgently summoned home by his father, his sudden disappearance without notice or excuse set "tongues wagging". After the war he reappeared and resumed his attachment with the Agricultural Department to complete his study. Leonard Kent was the owner of Mt. Rich Estate. People in the area thought that his wife was German, and she could go nowhere without being watched by the populace. Although the overseer to Marli Estate, Mr Sherwin, was English, for some reason he was thought to be German, and a spy. His every move, too, was watched by the community.

German sympathisers were regarded as only slightly less reprehensible than spies. Some of these sympathisers spread anti-British, pro-German and anti-war propaganda in the Caribbean, hoping to sow mistrust and antagonism for Britain. They hoped to spread admiration and support for Germany among disgruntled people in Grenada. Cecil Edwards remembers an old, brown-skinned man with curly hair who lived at the corner of the Mt. Moritz Road and Grand Mal. It was believed that this man had come from Barbados. This man seemed to have great knowledge of German troop movements in Africa and lectured the older men in the village on the likely events of the war. He was certain that the Germans would crush the British in Africa, and eventually win the entire war. But no one troubled him, and he lived out his life, dying in the 1960s.

Owen Wells from Grenville was not so lucky as to be listened to and left alone. Nicknamed "The Master", Wells and his wife, affectionately known as "Mother Wells", were very popular citizens of Grenville. Wells was one of the Grenadians who followed the events of war meticulously and, like many others, began to fear that Hitler would be victorious. Sometime in 1942, he foolishly voiced his opinion while socialising in a bar near the Grenville Police Station. He was heard to say that people should begin to learn German and prepare for Hitler's domination of the world, including Grenada. Some listeners took exception to his point of view and reported his utterances to the police, who came immediately and arrested him for sedition. This caused an immediate uproar in Grenville, because the majority of the population thought that the police action was extreme, and certainly did not think Wells was subversive. Moreover, at this stage in the war, many had the same fears and thought as Wells did. It was just that they did not voice these fears in public.

Many citizens converged on the police station protesting at Wells' arrest and seeking his release. The police in Grenville, concerned at the commotion, referred the matter of the growing protest to the Chief of Police in St. George's by telephone, and in short order the Chief arrived and took Wells away by car to the Richmond Hill Prisons in St. George's. A speedy trial was arranged, during which several people were courageous enough to testify as character witnesses for Wells. Although he was found guilty of uttering subversive statements, Wells was quickly given a pardon by the Chief Justice of the Leeward and Windward Islands, Clement Malone [48], and freed. The incident was handled with "kid gloves" throughout, because the authorities were afraid of a riot among the population.

XI: NEWS AND BROADCASTING

The BBC had transmitters at strategic points in the Caribbean enabling their programmes to be picked up in Grenada and other Caribbean countries, and radio reception had begun in Grenada in the 1930s. However, in 1939, still only a very few owned their own wireless sets, and reception was often very poor. The static cracked and spat, causing listeners to "cock their ears" in studied concentration to make sense of the broadcast. News bulletins from the BBC were relayed daily at 7 a.m. and 7 p.m. These were normally listened to by radio owners, but never so avidly as during the war for information on what was happening in England and the general progress of the war. Radio owners, especially in the rural areas, were generous and hospitable in allowing people from the surrounding areas to listen to their radios, and often positioned their radios so as to make listening easier for the crowd. Radios were also located in the offices of the District Boards, police stations, post offices and rum shops. Often, people would travel long distances to gather at the nearest accessible radio to listen to

---

48 Malone, later Sir Clement Malone, Kt., O.B.E., an Antiguan, was the first coloured man to serve in this post.

bulletins and programmes. When the broadcast was over, the crowd of listeners would discuss what they had heard, sometimes lubricating the discussion with rum, or making the long journey home short with lively commentaries. As a nightly follow-up to the broadcast, some men would gather at one home or another to discuss the news in whispers. The news would also be recounted in detail to those who had missed the broadcast. Generally, women were not part of the crowd of listeners at the District Boards and elsewhere, and neither were they informed of what the men discussed among themselves, as the men wished to shield the women from the harsh realities of the war. Mrs. Sylvia Greasley and her daughter Hermione were among the few women who were nightly listeners to the BBC news. Hermione remembers being taken by her mother as a child of nine to listen with the nightly crowd to the war news on the radio placed on a specially built shelf to hold the radio outside the offices of the District Board in Gouyave. Judith Parke and her aunt were another two of the exceptional women who followed the war news nightly. They would go together to listen to the nightly BBC news at the home of an old lady in Richmond Hill. As the radio reception was not always good, and listeners had to be very quiet to hear anything at all. Children present were cautioned not to make a sound during the broadcasts.

In the rural areas where there was no electricity, radios were even scarcer. Carlyle John remembers a *Delco* radio that was owned by a Lucas Andrew, a Grenadian recently returned from the United States of America. He lived about two miles from Carlyle's home at Snell Hall in St. Patrick's. It was a big radio, powered by motor car batteries. It had an extensive antenna mounted on a bamboo pole on top of the house. Mr. Andrews kindly allowed the local community to congregate in his yard to listen to the 7 p.m. BBC news bulletin, and would graciously turn the volume up so everyone could hear. At least one or two people went from Snell Hall every night to listen, and returned to tell the others what news they had heard. Other people in St. Patrick's with radios were Mr. Gill, Mr. Alister Glean, Mr. Gordon Gentle, and Mr. George Kent.

This practice was repeated throughout Grenada. Harry Ogilvie who lived in La Digue, St. Andrew's, remembers going with the older men to listen to Annie Hughes' radio at Prospect. People living within walking distance from the Noels' home at Carlton Junction would also walk there and congregate in the road to listen to the Noels' radio.

Bertrand Pitt remembers that Mr. Noel would play the radio very loud so that the people standing in the road could hear. Other radios in the Grenville area were at Gordon Hutchinson's rum shop, and at the homes of Mr. Rupert Rapier and Mr. M. Z Mark. Besides *Delco*, other popular brands of radios were *RCA Victor*, *Bush* and *Pye*, all sporting extensive aerials, and run off motor car batteries in the areas that were then not electrified.

It was a common practice to reprint some of the content of the BBC programmes in the newspapers, with editorial commentary. On 23rd May, 1943, *The West Indian* newspaper reported that:

> Sergeant-Pilot Julian Marryshow participated in Monday night's BBC special Empire Day programme. In a short talk he spoke of the way in which Grenada sought to help the war effort by subscribing to various funds both for the supply of weapons of war and for comforts for the fighting men. It was very appropriate that Sergeant-Pilot Julian Marryshow should have been selected to speak for Grenada for although he received his training in the Trinidad Air Scheme, financial arrangements had not yet been made for the dispatch of Grenada RAF Cadets to England, and it was the Grenada War Purposes Committee which stood security for his training expenses and voyage overseas. The Committee's gesture has been amply rewarded by Marryshow's fine showing as a fighter-pilot since earning his "Wings".

However avidly the BBC news bulletins were followed, neither the BBC nor *The West Indian* newspaper carried the type of war news that directly affected Grenada. There was the war "out there", as reported by the BBC, and there was the war as it affected Grenada, which, like the news of what was going on in and around Grenada, was transmitted only by word-of-mouth from man to man. Rum shops and gentlemen's clubs were popular venues for passing on information about the dangers to schooners, their passengers, captain, crew and cargo, the many torpedoed ships and passengers on them known to Grenadians, the merchants who had lost shipments of goods to enemy action, survivors who had reached Grenadian shores from torpedoed ships, and injury or death of West Indians serving in the armed services abroad, whose families were known in Grenada. Sightings of U-boats near Grenada were discussed as well as the damage they were doing to shipping. No one really expected that the sinking of a schooner or the sound of the distinctive engines of the U-boat near Grenadian shores would make the next BBC news bulletin. Thus the dangerous activity all around Grenada was not generally known. Only selected circles like the businessmen on the one hand, and sea captains and their crewmen and those they chose to speak with on the other, were really aware that the sea that placidly lapped the shores of their beloved island was now fraught with danger.

About half-way through the war, Germany began operating a radio station in the Caribbean called "Debunk". The objective was to broadcast propaganda and entertainment to the American troops in Trinidad while they idled in readiness for action. This station hoped to broadcast the sorts of things that would foster disunity between troops of various nationalities and ethnic backgrounds, and between the troops and the local population. To counter this German propaganda, the USA decided to set up a branch of the USA Armed Forces radio network in Trinidad. Primarily it was hoped that these radio broadcasts would be the choice of the troops, not only helping to relieve boredom but also keeping them in touch with "home". Secondly, it was hoped that these broadcasts would foster better internal relations amongst the troops, and between the military and the people of Trinidad. The station, WVDI, was opened in May 1943 and broadcast not only to Trinidad but to the southern Caribbean as well. This was another source of entertainment and war news for the people of Grenada who had access to radios. WVDI broadcasts continued to be enjoyed long after the war ended.

The practice of reading the newspapers aloud and giving verbal renditions of the news was commonplace throughout Grenada, the rum shops being a favourite place where this was done. As a teenager, Cosmos Cape was enlisted by the patrons of Hutchinson's rum shop to read and interpret the happenings of the war for them. In preparation for a session, he delved into his atlas and drew maps so he could give the patrons a visual image of the troop movements and engagements as reported by the BBC. As a reward for his illustrated war "lectures", Mr. Hutchinson fed young Cosmos Cape buns, with sweet drinks to wash these down. Official news bulletins were posted at the post office and police stations, and again, one person would read these aloud for the benefit of those standing with him who could not read themselves.

The West Indian newspaper served the Grenadian public well as an important source of news. It came out six times a week and printed information received from Britain as well as stories of local interest. Carlyle John remembers that *The West Indian* newspaper was shown around in Snell Hall, read aloud to those who could not read, or a digest of its contents disseminated to those interested in keeping abreast of what was happening but who did not have direct access to the newspaper. Oris Teka remembers reading *The West Indian* to her father on each day that it came out. The *Trinidad Guardian* was also a valued source of information and was regularly received by a lucky few in Grenada. Lennard McLeish, Pat McLeish's uncle, regularly received this newspaper and passed around the copies when he had finished with them.

XII: FINANCIAL HARDSHIP AND MIGRATION

A fortunate few profited financially as a result of the war. When Pearl Harbour was bombed on 7th December 1941, it was quickly discerned that there would be no more shipments of nutmegs from the Far East to Europe or America for a long time. This meant that the price of Grenadian nutmegs would rise. Those who could afford it quickly bought as many nutmegs as they could. Even though the shipments of nutmegs from Grenada stopped shortly after this due to the dangers posed by the U-boats, dealers who had nutmeg stocks were to make fortunes after the war.

Generally, however, during the war people had to economise and make their money stretch. No Grenadians starved, but amongst the very poor finances were sometimes so bad that, on occasion, family members went hungry. When there was not enough food, the youngest children got no breakfast. They were kept home from school, and later in the morning were given sugar-water to keep up their energy. Grenadians responded to the poverty and scarcity of food with their natural generosity. They shared what food they had with those that had nothing to eat, even though it meant less for them and their family.

At almost every level of society, people tried to see how they could bring in additional income. Women made delicacies such as crab backs for sale. Some made hats and handbags out of local material, both to replace their worn out apparel and to sell. Judith Parke's aunt used to make hats and bags out of straw and *lapit*, a tough succulent plant, whose leaves, when stripped and dried, produce strips of long-lasting filaments that can be used for weaving. Dora Mitchell utilised white pine for the hats that she made. Children could earn one penny a yard for plaiting the material for hats into strips, which would then be sewn together to make these accessories for ladies.

As the war progressed, small farmers who had fields of cocoa could not sell their crop to the larger planters. Cocoa was no longer being shipped due the danger of losing the shipment and all its value if the vessel were a victim to a U-boat. There was no shipping insurance during the war, and after 1942, the majority of shipments of cocoa and nutmegs — some estimates run as high as nine in every ten shipments — were lost in transit to the USA and the UK. As planters would only get paid when shipments were received at their destination, several got in debt to the bank or to agents such as Hubbard's and Hankey's when their shipments went to the bottom of the sea. As a result of this, many planters went into receivership and lost their estates. Shipments of cocoa and nutmegs eventually ceased until the danger from U-boats abated.

The curtailment of shipments of cocoa and nutmeg was also devastating to the Grenadian small farmer because an important source of income was no longer available when merchants stopped purchasing and exporting. Bags of cocoa remained in a corner of many small farmers' houses for the rest of the war. Without the shipments of cocoa and nutmegs, planters were also suddenly unable to provide work to sustain their labour force, and many estate workers found themselves unemployed.

This dire situation of many poor people was relieved when opportunities for employment opened up in Trinidad, Aruba, Curaçao and Venezuela, and men reluctantly left their families for countries to the south. Most stayed just long enough to make a little money and returned to Grenada as soon as they could. Some remained, most sending for their families, adding to the number of persons of Grenadian origin who migrated to other parts of the Caribbean and South America since the eighteenth century. Some members of the middle class also migrated due to the lack of opportunities at home. This was particularly true of the educated black middle class, because a large degree of patronage still existed in

Grenada — a society not yet quite emerged from the restrictive ascriptive plantation stratification. Educated blacks found it difficult to get a job commensurate with their qualifications, because office jobs and entry positions for middle management in the civil service were still preferably given to the white and light-skinned applicants from families that were known to the employer. In spite of the dangers to local shipping posed by the submarines, the war years saw the peak of Grenadian inter-island migration to the islands where there was well-paid work to be found. During 1941 to 1944, the net loss of the population of Grenada to migration was 8,300 persons.[49] Crowds would gather outside the shipping offices waiting for the opening hour when they could buy tickets. Four of the many boats that plied regularly between Grenada and Trinidad carrying migrants were the *Island Queen*, the *Enterprise S.*, the *Rose Marie* and the *May I Pick*.

In Trinidad, there was the need for all types of labour. As soon as the Americans began building the bases and roads in Trinidad in 1941, Trinidadians abandoned agricultural work for better-paid jobs in construction and other jobs with the Americans. Anyone who could drive a vehicle was also sure to get a well-paid job. It is estimated that some 40,000 persons [50] were employed in building the bases in Trinidad and with other American projects, such as construction of the Churchill Roosevelt Highway and working on the docks. When these were finished, 9,000 local civilians were to find continuing work on the Chaguaramas Naval Base, alongside 3,000 Americans. Trinidad alone could not supply this amount of workers, so there was an acute shortage of all sorts of labour in Trinidad. Citizens of the Eastern Caribbean flooded in to fill the void.

The Agricultural Society in Trinidad was successful in pressing the Government to relax some of its immigration laws in order to bring in workers from the small islands to maintain the sugar production and to replace the bus and transport drivers who had left their regular jobs to satisfy the employment need of the American military. The depletion of agricultural labour was especially worrying because there was a dire need to fulfil the demands of the *Grow More Food* campaign. Like Grenada and most of the Caribbean islands, much of the food eaten in

---

49 Harewood, (1960) P. 66.
50 Fullberg-Stolberg, P. 97

Trinidad was imported, and when the supply of imported food was cut off as a result of the war, Trinidad was faced with the situation of having to feed herself. Some migrants did go into agriculture as planned, but others worked at anything they could get to do. A Grenadian seeking work in Trinidad at this time writes that he worked

> First as a helper in a Chinese shop named Marley & Company Ltd. Later I worked in a cocoa station for one dollar a day. Next as a helper for an East Indian on his truck, transporting coals from Sangre Grande to Port-of-Spain...[51]

Some migrants were legal, recruited by the Agricultural Society in Trinidad. Others just arrived. Irie Francis, in his biography, makes mention of the "Rat Passage" that was used by illegal migrants to get to Trinidad. This entailed boarding a schooner or sloop under the cover of darkness and hiding among the goats and sheep that were being transported to Trinidad to be sold. When the vessel docked in Trinidad in the wee hours of the morning, each passenger went his separate way, undetected by the authorities. Other sloops and schooners put into deserted bays in the rural parts of Trinidad to give their passengers the best chance of an undetected arrival.

Problems arose for the migrants when the building of the bases was complete and Trinidadians began searching for new jobs. Trinidadians began to feel that the "small islanders" were depriving them of jobs, food and resources made scarce by the war, and were a threat to the Trinidadians themselves by swamping the local population. One writer sees this period as being the genesis of the Trinidadian prejudice against "small islanders",[52] and quotes one of Lord Invader's calypsos which sums up the local feeling:

> No flour, no rice in the land — believe me, too much small island.
> Yes, they come by de one, de two, de three
> Eating our food and den leaving us hungry
> Dey will send fuh their brother, dey aunt, and dey sister,
> Dey cousin and also grandmother
> So small island go back wey yuh really come from.

The hostility became so palpable that around April 1942 many of the small islanders returned home prematurely, feeling they were now

---

51 Francis, P. 19
52 Gopaul-Maragh, P. 42

unwelcome in Trinidad. The Trinidad Government passed Ordinances in 1942 and 1944 to prohibit immigration to Trinidad, except for persons under contract to perform agricultural labour. However, many immigrants remained, and new migrants continued to arrive.

Among the migrants were several educated young men of the coloured middle class, quite willing to engage in manual work because the pay was higher than anything they could hope to earn in Grenada as clerks and junior white collar workers. Among the migrants to Trinidad were Alister Hughes, Sydney Steele and Elton George Griffith. Elton George Griffith left Grenada around 20th June, 1941, with plans to leave from there for Syria. He decided to stay in Trinidad, involving himself in the struggle to restore the Shouter Baptists' rights to their form of worship.[53] Alister Hughes, who returned to Grenada after the war to become a noted auctioneer and journalist, worked first as a stevedore and then as a supervisor on the Trinidadian docks.

The story of John Watts is an instructive one. John Watts was from Sauteurs, but like so many students, he boarded in St. George's when he began to attend the GBSS. After graduating from GBSS he was employed as a teacher at the River Sallee Government School in St. Patrick's. He was encouraged by his friends to leave this low-paying teaching job to seek better wages in Trinidad. His first thought was to seek a job with the Americans, and he was offered a job to clear virgin forest for the construction to follow. However, he was unaccustomed to manual work, and disappointed that this was all that was available. He went to visit a friend, Ignatius Wilson from Sauteurs, who was at the time the Chief Secretary to the Manager of the Engineering Department of United British Oil Company in Point Fortin. Wilson knew that his company badly needed educated clerks, and he arranged an interview for Watts the next day. Watts was hired on the spot and immediately started a well-paying and interesting office job, with excellent working conditions. Watts worked for the Oil Company until 1946 when he left for the United States of America to study dentistry. Watts returned to Grenada some years later, and quickly became prominent in his profession. He also played a role in the politics of the country, founding the Grenada National Party. He was knighted by Queen Elizabeth II.

---

53   See Jacobs, (1996).

The oil companies in Aruba and Curaçao expanded rapidly at this time to meet the requirements for oil and oil products for the ongoing war. Grenadians were among the "small islanders" who migrated to Curaçao and Aruba for jobs in the oil refining industry. Eric Matthew Gairy and Herbert Blaize, who both returned to Grenada as politicians and later became prime ministers, were among the Grenadian migrants to Aruba. Rupert Bishop and his wife, Alimenta, also spent their early married life in Aruba. Alimenta returned to Grenada for the birth of her first two children, but by the time her third child Maurice was to be born, it was far too dangerous to return, as this was the time of the height of the U-boat activity. Therefore, Maurice Bishop, their famous son, was one of the many Grenadians who had an Aruban birth certificate because his parents had to seek work away from home at this crucial time.

XIII: THE CHILDREN DURING WARTIME

A word must be said about the children of Grenada who had to live through the war with its many frightening aspects. Most adults tried to shield the children from disturbing news and thoughts occasioned by the war, but these efforts were neither entirely successful nor universally practiced. It was difficult to hide the fact of the war from children who had gained a sense of their surroundings. Children inadvertently overheard their parents' conversations about the war, and overheard the radio transmissions.

While some adults tried to shield the younger ones from the news, many small children knew all the details. Alister McIntyre's father Meredith "Merry" McIntyre was an avid listener to the BBC, and so the declaration of war did not come as a surprise. Alister, who was then 7 years old, remembers his father calling him and telling him that war had been declared. His father told him that there was a man called Hitler who had been doing some quite unacceptable things in Eastern Europe and to the Jews. Churchill had decided to declare war to stop Hitler. To the young Alister, this boiled down to "Hitler was a bad man and Churchill was their saviour". Alister commandeered his four-year old brother for an army of two parading about their yard with long pieces of wood for rifles.

Some older children were allowed to listen with their parents to the BBC news and to discuss the war news with them. Children were also told war news by the principals of their school. Enid Bain [54], whose family was from Upper Capital, St. Andrew's, and who attended the Birchgrove R. C. School, remembers that Fr. Godfrey Austruther O.P. would come and give the children a lecture on the war every morning. Unknown to the parents, some of the older children discussed the war with the younger sisters, brothers and friends whose parents had assiduously tried to keep in ignorance. However, war news seemed to many children tales of far off and distant lands and almost like a "Nancy Story".

At first, there was very little evidence of war in Grenada, but when unusual activities of the war such as blackouts were instituted, some children hated the lack of light in the house after dark and having to go to sleep in the utter darkness. Others thought the blackouts were "fun", mostly because their mothers expended so much effort and imagination to make them so.

Nevertheless, some children picked up anxiety and worry from their elders and, with little idea of geography, could work themselves into a panic. Bertrand Pitt was usually sent by his parents in Grenville to spend the long summer vacations with relatives in St. David's. The first time he went to St. David's after the war started he witnessed youths going up into the hills to cut sticks for pretend guns. He remembers them marching along with these "guns" to British marching songs. In his simple child's mind, Germany was "just up the road". When it was time to return to Grenville, he was afraid to leave St. David's because he was terrified of meeting "Hitler by Hope". [55]

Carlyle John remembers the stories of the U-boats, and the fright of the local people of the stories of them. He remembers being scared at these stories of the war and wondering what a gun or a bomb looked like. He also remembers hearing that local people had gone to join the army. But the full understanding and impact of the war and what it meant was spared him.

In some cases, boys were sometimes allowed to tag along behind their fathers, seeing what they saw, and hearing what they heard. Moreover,

---

54   Mrs. Alistair Charles
55   Hope is a village in St. Andrew's, along the road to Grenville.

in the audience were exposed to all the horrors and destruction of the war. The general opinion among the children was that the Germans were bad. Janice Bain has a vivid memory of one of her friends saying that Germans did not like black people so, if the Germans conquered Grenada, Grenadians "would be in for a whole lot of trouble".

The first time he saw a plane flying over Sauteurs, Carlyle John remembers that he was so frightened that he went to hide. At nights when they were trying to sleep, Margaret Phillip [56] and Geraldine Sobers used to hear aircraft droning overhead. Like so many other children, the sound of the aircraft frightened them; they did not know if it was friend or foe. Would this be the night when aircraft would drop bombs on them and their families? Geraldine Sobers remembers her mother telling them as children not to show any lights when they heard planes overhead, as any light at all would show the Germans where the houses were. She also urged them to be very quiet lest the Germans heard the noise or whimpers. Parents quietened their children's fears, often all the while as scared as the children themselves. Many did not know the difference between the sound of German and American aircraft, and did not realise that a German aircraft did not have the range to fly to the Caribbean and drop their bombs on Grenada! Sometimes the noise of the planes seemed to go on forever.

The first view the children had of a zeppelin was exceedingly traumatic. They had never even seen a picture of a craft such as this before. They were totally unprepared for its arrival, and in many instances the first encounters were fraught with pure terror. Judith Parke revels in hindsight at the memory of her first sight of a zeppelin. The reading class for her group at the Morne Jaloux Roman Catholic School was taking place in the patio of the school, as was often the case on nice days. Raising her eyes from her book, she noticed something like an egg in the sky. For a while she kept one eye on it and one eye on the reading lesson, growing more apprehensive as the object grew larger and larger, and more like a pumpkin than an egg. Finally, she could no longer contain herself and burst, "Sir! Look at a pumpkin the sky!" The teacher reacted in a much less calm and controlled manner than little Judith and pandemonium broke out, with children and teachers rushing everywhere, some crying

---

56   Mrs. Reginald Dowe

out and some trying to hide. Both children and the teachers thought the Germans had come to kill them. Although the zeppelin disappeared from view without any harm coming to anyone, that was the end of school for that day.

Oris Teka and her sisters had a similar fright when, for the first time, one passed within sight of where they lived in Woolwich Road. They began to cry in distress, believing that their end had come. When Elinor Salhab[57] first saw a zeppelin, she thought it was a flying German bomb.

As the sight of the blue-grey zeppelins became common, the fear of these craft diminished. By the time Hermione Greasley and Paul Scoon saw zeppelins from their homes in Gouyave, news had spread that zeppelins were craft of the Navy of the United States of America and, as such, were friendly.

From the hill communities of Grenada, great expanses of the sea can be viewed. Several children remember seeing at night the horrible sight of boats on fire out at sea. Judith Parke grew up in the Richmond Hill community above St. George's. From her home, she could see the convoys of oil tankers and other ships in convoy on the way from Trinidad to Europe. She understood that convoys were a precaution against enemy attack and the loss of life and shipping.

Children suffered when they were deprived of their fathers or other members of their families when the adults in their lives joined the various armed forces or migrated to find jobs in Trinidad or Aruba.

The children of those years, now adults, admit that although they might have been frightened in response to certain happenings, they were not constantly depressed or worried during the war. Parents did their best to shield children from deprivations and scarcities and provided for them, so that many never knew real hardship.

Life during the war had its light moments as well. Many pets at this time were named for prominent generals. such as Timoshenko. One family living in St. Paul's acquired a ferocious black and white dog, who the father of the family decided should be named Rommel. For the most part, life in Grenada, at least for the children and young adults, continued almost as normal. Where there were concerns, the day-to-day

---

57 Mrs. Gordon Lashley

distractions of childhood allowed the war to be temporarily forgotten. Confidence in their own existence also grew, so that they could cope with the worst experiences and stories. Although conscious of the war, the childhood of many was filled with the usual things such as minding animals, housework, school, good friends, long walks and, in the middle class, birthday parties and music lessons. So was the childhood of all Grenada's children filled with the usual rather than the unusual.

Map of Grenada
Copyright Xandra Fisher (the artist) 1981.

Map of the Caribbean
Copyright Xandra Fisher (the artist) 1981.

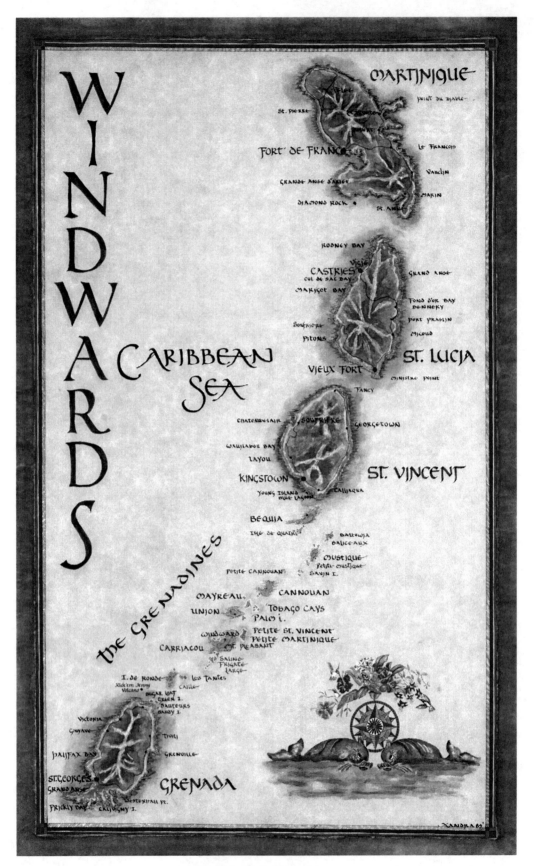

Map of the Windward Islands
Copyright Xandra Fisher (the artist) 1984

Map of Trinidad and Tobago
Copyright Xandra Fisher (the artist) 2007

Sir Henry Bradshaw Popham K.C.M.G.,
Governor and Commander-in-Chief of the
Windward Islands (including Grenada)
1937-1942

Sir Arthur Grimble K.C.M.G., M.A.
Governor and Commander-in-Chief of the
Windward Islands (including Grenada)
1942-1946

Terrence B. Comissiong, the Colonial Secretary in 1944
as a young soldier in World War I
(Picture courtesy Geoffrey Comissiong)

Almost the complete horseshoe shape of the St. George's Harbour can be seen in this photo. The fire station with its prominent tower can clearly be seen. The *Island Queen* and *Providence Mark* departed from a mooring just opposite to this buildings on 5 August, 1944. Photo: Windward Islands Annual 1963

Another view of St. George's Harbour taken in the 1960s, showing buildings on the side opposite to the fire station. The schooner is about to round Hospital Point. The doctor's house where the Slingers once lived can just be glimpsed above Marine Villa, which is situated almost at the end of the promontory and which was the residence of the Chief of Police. Photo: Windward Island Annual 1963 through the courtesy of the Chronicle of the West India Committee.

A view of St. George's Harbour prior to 1955 which includes the Lagoon. The ballast ground where the *Island Queen* was built is the little piece of ground jutting into the Lagoon on the right, seemingly stretching to meet the old pier and warehouses destroyed by Hurricane Janet. The fire station with its tower can also be seen. Photo: Windward Islands Annual 1958-59 through the courtesy of "The Team"; house journal of the Cubitt Group.

Everybody's Store at the Market Square in St. George's in the 1940s. Photo: Leo Cromwell

Buses in Market Square in the 1940's. Photo: Leo Cromwell

Bruce Street in the 1940s.
Photo: Windward Islands Annual 1964

View of Granby Street in the 1940s.
The three-storey corner building was known as "Masanto's" and it was here that Oswald Callendar had his photo studio. Photo: Windward Islands Annual 1957

World War II Barracks, Tanteen
View of barracks constructed to house the members of the Southern Defense Force. The site had been prepared for the use of the Grenada Boys' Secondary School, and several years after the War, the school finally moved to this site. Some of the Barracks were used until 2004 when Hurricane Ivan damaged them beyond repair.
Photo: Leo Cromwell

Schooner with wartime markings W546 (W for Windward Islands).
Photo: Windward Islands Annual 1956

Launch being shipped to Grenada from Port of Spain Docks bearing WWII markings T16
(T for Trinidad). Photo: Windward Islands Annual 1956

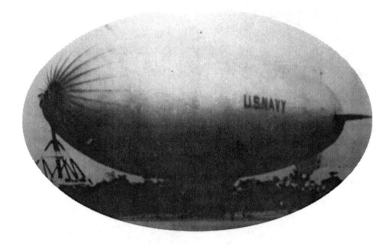

Zeppelin belonging to the U.S. Navy at Carlsen's Field in Trinidad.
The use of blimps was crucial to spotting German U-boats.
Photo: Trinidad National Archives

Herbert Payne, an officer in the Southern Defence Force, on his motorcycle at Ross's Point with a young attendant. Ross's Point and Tanteen were stations for the Southern Defence force during the War. With the rationing of fuel, motorcycles and bicycles were commonly used for transport.
Photo courtesy Nellie Payne

Tomb of the Unknown Sailor, Hillsborough Cemetery, Carriacou.
Inscription reads: A sailor of the 1939-1945 War/ Merchant Navy/ Buried 21st July 1942/ Known unto God.

Cup presented in gratitude for the help received from Grenada from 21 survivors of the Norwegian Vessel *Baghdad* sunk near Grenada on 3rd June 1942. The Cup is now on exhibit in the Westerhall Museum, St. David's, Grenada.

Many West Indians served in the Royal Air Foce during World War II. Picture shows an Air Crew training at Peterborough. Grenadian Pilot Julian Marryshow on the extreme right.
Photo courtesy 602 (City of Glasgow) Squadron Museum, Scotland.

Julian Marryshow was one of the Grenadian pilots who flew night bombers. Picture shows the 'B' Flt. Sumburgh November 1942. Pilot Julian Marryshow is on extreme right.
Photo courtesy of the 602 (City of Glasgow) Squadron Museum, Scotland.

Launching of the *Island Queen* at the Ballast Ground, St. George's, 1938.
Photo: Alister Hughes

Another photo of the launching of the *Island Queen* at the Ballast Ground, St. George's, 1938.
Photo: Alister Hughes. Enhancement by Modern Photo Studio.

Chykra Salhab and his brother André on board the *Island Queen*. Photo courtesy Elinor Lashley.

Another photo on board the *Island Queen* with Osborne Steele, Chykra and André Salhab.
Photo courtesy Elinor Lashley

Heavily laden trading schooner similar in size to the *Island Queen* entering St. George's.
Photo: Windward Islands Annual 1962.

FLASH! FLASH!!

Excursion to St. Vincent
By Auxiliary Schooner
"ISLAND QUEEN"

Contact:

GORDON A. CAMPBELL,
at 4 Halifax Street, St. George's,
on or before Saturday July 29th, 1944,

Advertisement for the Excursion to St. Vincent in *The West Indian* Newspaper

The former Scoon Guest House on St. John's Street

Salhab home and complex on Scott Street, now in disrepair

The Harbour in Kingstown, St. Vincent has changed vastly in the years since the disappearance of the *Island Queen*. The area where the *Island Queen* and other inter-island schooners docked in the 1940s is now occupied by the new cruise ship terminal. In this photo, Dr. Edgar Adams looks at a section of the harbour near to the old docking area, where the mail boat for the Grenadines and other small vessels now dock.

Magistrate Henry Steele, one of the two Commissioners on the Enquiry into the loss of the *Island Queen*.
Photo courtesy Valerie Steele.

Iris Rowley and Family. Iris Rowley was forced to acknowledge her husband's death before she was ready in order to collect the gratuity paid to civil servants who died on the *Island Queen*. Photo courtesy Robert Rowley.

Sydney Wells, famous Grenadian schooner captain (with sister). Wells led the early search for the *Island Queen* as well as the search along the Venezuelan coast. Photo courtesy Brenda Williams.

T.A. Marryshow, Grenadian Statesman and Editor of *The West Indian* Newspaper who lost three children in the Disaster.

PART II

THE CARIBBEAN THEATRE OF THE WAR

 **I: AMERICAN NEUTRALITY AND THE WAR**

Great Britain and France entered the war with the lofty and worthy ideals that they would defend the sovereignty of the European states threatened with invasion to ensure the balance of power in Europe. To their great surprise and shock, Germany overran France in June 1940, leaving England alone to defend herself against Germany and her Allies. Therefore Britain's first priority was to apply these ideals to Britain herself, as the successful invasion of France meant that Britain now faced the very real possibility of invasion of her shores, and of a German conquest. For two years, Britain fought for her very existence. There were times when the possibility of invasion and defeat was very real, so much so that there was an order to remove all signposts, so that it would not be easy for the Germans to find their way around Britain. Apart from the humiliation of defeat, the thought of living under a totalitarian regime was anathema to Britain.

In the months leading up to the war, and in the first months of the war, powerful factions in the United States of America were determined that the USA was to follow an isolationist policy with regards to the war. Their slogan was "America First". However, the war had to be of interest to the USA, because should Germany and her Allies be victorious, Germany would be at the doorstep of the USA. That was not at all desirable. The USA therefore modified her isolationist policy to allow for

- Safeguarding the Panama Canal — the vital shipping channel that linked the Atlantic and Pacific Oceans. This link, created to speed up commercial shipping, would be vital to the exigencies of a global war, when men and material had to be moved across the globe with speed and efficiency.

- Preventing of the oil and bauxite from the Caribbean region from falling into German hands. Before the war, Germany was the largest purchaser of oil from Aruba, Curaçao and Venezuela. At the outbreak of the war, American oil companies in these countries ceased selling to Germany because the oil would be needed by Britain. The USA saw that it was in America's best interest to ensure that sufficient oil to supply the needs of the war reached Britain.

- Preserving the USA's extensive trade with South America. Bauxite from the Guianas was vital to the USA's expanding aircraft industry, while sugar, coffee, fruits, leather and beef were imported in considerable quantities.

In addition, from the outset of the war, there was constant dialogue between Britain's First Lord of the Admiralty, Winston Churchill [58], and the President of the USA, Franklin Roosevelt, as to how the United States could help Britain and France and at the same time maintain neutrality in view of the necessity to protect itself and its interest in the Americas. The USA began negotiating with Britain for permission to set up US military bases on British Caribbean territories for the protection of the Americas and Caribbean Area. US personnel were sent to the British West Indian Islands to identify likely sites for these bases. The USA also spearheaded *The Declaration of Panama*, which was signed on 3rd October, 1939 by 21 American republics, including the United States.

---

58 Following the resignation of Neville Chamberlain on 10th May 1940, Winston Churchill became Prime Minister of the United Kingdom.

This declaration, signed one month after Britain declared war against Germany, mapped out a Pan-American Neutrality Zone that included the Caribbean. The document was a statement of intent that the region would remain neutral, and therefore any attack on the countries who signed the agreement would be breaching this protocol, and would be considered an act of aggression.

When Germany invaded France in June of 1940 and, in a matter of weeks, achieved the defeat of that country, all the fears the USA had had regarding her own safety surfaced, and she had to reconsider her position of neutrality in light of the changed situation in Europe. The much-discussed and negotiated Anglo-American Naval and Air Base Lease Agreement, or the *Land Lease Agreement* as it is commonly called, was signed on 27th March, 1941. In addition to the lease on certain property on the territories in the Caribbean that were part of the British Empire, on which would be constructed United States air and naval bases, the terms of the *Land Lease Agreement* included that the USA would be responsible for the defence of the area formerly covered by the *Declaration of Panama*. The USA would also immediately supply Britain with 50 old but serviceable American destroyers from World War I to act as protection for the convoys of merchant ships going to Britain, and thus help to keep Britain's shipping routes from the Caribbean open. The U-boats could travel much faster than a convoy, which had to travel at the pace of the slowest ship. U-boats were also equipped with devices to pick up the sound of a convoy, enabling the U-boats operating in "Wolf Packs" to attack. In this event, the convoy would not stop, but one ship near the rear of the convoy would be designated to take the risk of slowing to pick up survivors. In spite of the danger, convoys did provide some protection for the shipping that could not cease, as it was Britain's lifeline.

If her trade was disrupted, Britain would be crippled because she could not meet all her needs domestically, and needed to sustain her trade routes for her own survival. Above all, oil needed by the United Kingdom for its aircraft, ships, tanks, vehicles, factories, and other essential machinery of sophisticated modern motorised warfare came wholly from Aruba, Curaçao, Trinidad, Venezuela and Texas, USA. Britain had access to oil sources in the Far East and the Persian Gulf, but this oil was utilised in those theatres of war. In order to keep her war machine functioning, Britain needed four oil tankers from the Americas arriving at her oil terminals every day.

Britain was also dependent on the bauxite mined and refined in the Americas, and on food grown in this region to help feed the people of Great Britain and her military. There were other commodities used daily in Britain that were shipped from the United States and Canada through the Caribbean and across the Atlantic. The well-being of the colonies would also suffer if the trade links suffered any disruption. At all cost, the shipping lanes between the Caribbean and Britain had to stay open.

The USA had been surreptitiously sending Britain all the material aid she could without being deemed to have joined the war. By the end of June 1941, America had supplied Britain with US$75 million worth of goods free of cost. After the first bombing raid by Germany on London, America started mass-producing fighter planes for the British, and made available two million tons of cargo ships and oil tankers to Britain. The USA also relieved Britain of the responsibility for defending the Caribbean and the bauxite and oil transportation routes out of the Caribbean which Britain was unable to do because of the pressures of the war in Europe. The USA would dedicate some of its vast resources to

> (E)stablish air patrols for the protection of neutral shipping in the Caribbean Sea and in the approaches to the Panama Canal, and also in the east coast of South America. [59]

It was also agreed that the United States would patrol and defend a

> Zone that would be drawn elastically to cover not only Canada and Newfoundland but also all British, French, and Dutch possessions in the circum-Caribbean. Moreover the zone would extend down the whole west side of the Atlantic to include South America. [60]

In response to this agreement between America and Great Britain, Germany deemed America to have breached her own protocol. America was no longer neutral, but fighting an undeclared war. Around the middle of 1941, Germany began to sink American ships.

Plans were also made by Germany to surprise America with a simultaneous attack upon some of her overseas possessions. When America received the intelligence that the Hawaiian Islands, the

---

59 Baptiste (1988), P. 11
60 Ibid.

Philippines and other territories in the Far East were under threat of being captured by the Germany's ally Japan, America began to amass a great Pacific fleet stationed at Pearl Harbour on Hawaii, and a smaller fleet in Manila in the Philippines. On the quiet morning, 7th December, 1941, Japan attacked both Pearl Harbour in Hawaii, where most of the American Pacific fleet was at anchor, and Manila, where the other part of the American fleet rode at anchor. From this date, America joined the war, and Britain was heartened to be supported by a powerful ally.

II: THE MILITARISATION OF TRINIDAD

Despite the powerful isolationist lobby in the United States, those at the head of the Government knew that America could possibly be endangered by the war. Talks had been ongoing with Britain regarding the need to protect America and the Caribbean. Foremost among the proposals was the suggestion that the USA be allowed to build military bases in the Caribbean in the islands belonging to Britain. The proposals were accepted, and a team led by Admiral Greenslade completed the inspection of the territories of the British West Indies and identified sites suitable for the construction of military bases. Greenslade's team picked Trinidad as a major centre of American operations in the Caribbean because of the oil produced there, because of its proximity to the South American mainland and ability to refine Venezuelan oil, because of its strategic position astride some of the main shipping routes to and from South America and shipping heading to traverse the Panama Canal, and because the geographical features of the island were excellent for the purpose of creating a massive integrated military installation. By 28th December, 1940, the US military had aerially photographed and mapped Trinidad and sent in some American technicians. Sovereignty over the areas to become bases was transferred at a ceremony in Trinidad on 1st March, 1941, twenty-six days before the formal signing of the *Land Lease Agreement*. Present at the handing over ceremony were Major David Ogden, the District Engineer in charge of the construction of the base at Chaguaramas. With him were an advance guard of 60 marines and 10 soldiers. Soon this would swell to a force of 500 Americans and 5,000 local workmen. On lease for 99 years to the USA by virtue of the *Land Lease Agreement* would be 32 square miles of Trinidad soil, consisting of 11 square miles at Chaguaramas, 19 square miles at Cumuto, and small areas elsewhere for auxiliary airfields, supply wharves and recreational facilities.

At Chaguaramas, the Americans would build the biggest naval operating base and air naval station outside of the United States and one of the largest in the world. This base would occupy that entire peninsula and would include an extensive harbour, a Naval Air Station with a very large flying boat ramp and three hangars, accommodation for sixty flying boats, and a repair facility. Four hangars had been planned, but the parts for the structure of the fourth were lost when the ship carrying them was sunk by a German U-boat.

Soon US soldiers began to arrive in battle-dress and "armed to the teeth". Trinidadians saw for the first time jungle camouflage in all its forms, American jeeps and the monstrous Mack trucks. The entire island was quickly militarised. Apart from military installations, other activities of the war and the facilities to house them were scattered all over Port-of-Spain and the rest of Trinidad. Several large houses around the Savannah were rented or requisitioned for offices, and others became officers' clubs. Talbot, the American General in Trinidad, converted *Whitehall*, until then the private home of the Agostini family, into the headquarters of the United States Base command. Another mansion, *Daggerrock*, became a home for convalescent US sailors. George V Park was also leased to the Americans for the purpose of building barracks for American soldiers.

The American installations were built and became operational very quickly. Michael Anthony says of the building of the Base at Chaguaramas:

> Responding to their slogan "Time is short" the Americans (with their local workmen) seemed to get things done at incredible speed, for what seemed the work of months was accomplished in a few weeks. Soon, for all practical purposes, the naval base was complete, and shortly afterwards – on July 30 – Captain Arthur Radford arrived to take command. [61]

The base, built in an incredible six months, was formally commissioned on 1st August, 1941. Nothing was allowed to slow the pace of operations.

---

61   Anthony, P. 63

By this time, tyres were very hard to get in Trinidad, throwing the transportation system into chaos. To ensure that the local workmen got to their jobs at Chaguaramas, the Americans would collect the workers at designated assembly points in the morning in their huge Mack trucks. These trucks would also return the workers to these points after work in the evenings. When it was completed and occupied, the Naval Operating Base at Chaguaramas was a complete society within a society, with houses, recreations centres and commissaries. Over the period of the war approximately 100,000 US military personnel served in Trinidad.

Besides the enormous base at Chaguaramas, massive military, air and naval installations were constructed elsewhere in Trinidad, and supplementary land leases were made to the USA as the operations in Trinidad expanded. The Gulf of Paria was the largest protected harbour in the western hemisphere, and it surrounded both Port-of-Spain and Point-a-Pierre on three sides. Facilities were put in place to transform the entire Gulf of Paria into a large secure convoy centre and an anti-submarine base. The control of Port-of-Spain docks would be taken over by the US military, and were extended to enhance their capacity and effectiveness. Most of the war and construction material coming into Trinidad was landed on these docks in Port-of-Spain. Reclaimed lands, known as *Docksite*, together with lands along Wrightson Road resulting from the creation of a deep-water harbour, would be used by the American army as a storage depot for these materials, for the construction of offices and other buildings, and also for a camp for the Marine Corps. In addition, all other harbour facilities in Trinidad under US control were expanded and heavily fortified. The "Bocas" — or entrances to the Gulf — were mined, albeit against local advice that the extremely strong current would tear the mines loose. A Designed High Frequency Radio Direction Finder was installed at Chacachacare Island to get a fix on radio transmission of hostile aircraft or ships. By the time the USA entered the war, Trinidad had already been transformed into a virtual fortress, and the social and political atmosphere of Trinidad came close to that of an occupied country.

After America entered the war, Port-of-Spain Harbour and the Gulf of Paria were made both the terminus for the North Atlantic convoy route and the convoy assembly point for oil tankers going from the Caribbean oil refineries to North Africa and Europe. Ships met in the Gulf of Paria and were convoyed along different routes in and out of the Caribbean. By late 1942, the Gulf of Paria was also a training base used by carriers and planes belonging to the United States before they were dispatched to the Pacific theatre of the war via the Panama Canal. Final exercises for destroyer escorts and major warships also took place in the Gulf and, early in 1943, the training of the airmen and crews of aircraft carriers commenced. By mid-1943, the Gulf was seeing up to thirty convoys — approximately one thousand merchant ships — per month.

A large army base, built at Cumuto, was named Fort Reid for U.S. Major-General George Windle Reid, a World War I American military hero. It could accommodate a garrison of three divisions, but never exceeded twenty thousand troops [62], including two anti-aircraft regiments. A part of the extensive facility at Cumuto was an airfield known as Waller Field which could accommodate 100 aircraft. Waller Field was named for U.S. Major General Alfred J. Waller, who had been killed in a plane crash in 1937. Carlsen Field and Edinburgh Field were two other major airfields near Chaguanas. Carlsen Field was the site of the RAF training station set up to train pilots from the British Caribbean for service in Europe. Additionally, RAF aircraft were stationed at this field to assist the Americans and provide the Caribbean area with adequate air cover. The air bases at Waller Field and Carlsen Field were also home to submarine spotter aircraft, bombardment planes, and zeppelins.[63] Piarco, previously a civilian airfield used for light aircraft, was now greatly extended for military use, although it continued to be used by civilian aircraft. Airfields in Trinidad were often used as stop-overs for military aircraft earmarked for the war in Africa, such as planes for the Eighth Army in North Africa that were ferried through Trinidad. There were also smaller airfields elsewhere including at Mayaro, Toco and Crown Point in Tobago.

The waterfront at Carenage was deepened to create a landing for seaplanes, PBY "flying boats".[64] The Americans built a highway, which they diplomatically called the Churchill-Roosevelt Highway, to provide better access between the naval base at Chaguaramas and the army base at Cumuto. Another highway was built to Maracas Bay, ostensibly to make this bay available for sea bathing, since popular bathing spots at

---

62   Kelshall, P. 6

63   See Part IV Section III

64   The initials P.B. stand for "patrol bomber" and the Y was the identification letter for the manufacturer of this aircraft - the consolidated aircraft.

Chaguaramas were no longer accessible. The real purpose of this road, however, was to gain quick access to this strategic bay and the north coast. The new road to Maracas Bay was opened on 5th April, 1944.

Trinidad was made the point of entry into the Caribbean. Vessels and planes from South America had to be cleared at Trinidad before they were allowed to proceed to their North American or European destinations. Facilities were also set up in Trinidad that served Grenada and the Eastern Caribbean. Already mentioned as established in Trinidad for the Eastern and Southern Caribbean were the headquarters of the Southern Caribbean Force, and two prisoner-of-war camps. There was also a large censorship department in Trinidad, which was also the headquarters for the censorship activities in the Eastern and Southern Caribbean. This department also served as a cover for British and U.S. agents searching for Latin Americans who used the neutrality of Spain and Portugal to engage in smuggling or espionage for Germany.

III: THE MILITARISATION OF GRENADA AND OTHER EASTERN CARIBBEAN ISLANDS

Simultaneously, the islands of the Eastern Caribbean were militarised. As a part of the Bases Agreement, the USA was allowed to requisition land, and build and maintain facilities elsewhere in the Caribbean besides Trinidad. In the southern part of St. Lucia approximately one thousand acres around Vieux Fort were requisitioned for an airbase. Humfrey remarks that in a matter of weeks the Americans

> laid down an airfield complex capable of handling the massive silver B29's, which were eventually used to drop the atomic bombs on Hiroshima and Nagasaki; pushed a broad jetty out into the Caribbean; and almost as an afterthought, they set up all the buildings to house and administer 3000 men. The accommodation and the facilities made available to the airmen were unbelievably lavish by any but American standards. [65]

---

65 Humfrey, P. 165

This airfield complex was named Beane Field [66]. Among the facilities constructed were military buildings, roads, an enhanced water supply system, and a hospital. The gigantic runway complex featured large, long runways to accommodate the B29s and the famous Douglas B18 Bombers, nicknamed the *Flying Fortresses*. The nickname *Flying Fortress* was actually applied to both the B17 and B18 aircraft. The Caribbean has the distinction of giving the B18 Bomber its only war service, because all others were destroyed at Pearl Harbour.

From Beane Field the wide arc of sea passages covering the area between the Virgin Islands and Trinidad was patrolled. The obscure, sleepy village of Vieux Fort, which was situated close to the Base, became a prosperous boom town, with almost everyone working in some way for the Americans. Smaller bases were established at Vigie (with an airstrip), Reduit Bay and Cap. A large number of Barbadians and some other islanders, including Grenadians, helped to construct the St. Lucia installations, and a number of soldiers from Grenada were stationed there with the Southern Caribbean Force.

Antigua had two Americans installations — both in the north of the island. A naval sea plane station occupied the entire Crabbs Peninsula. An airfield, called Coolidge Airfield (after Calvin Coolidge, the 30th President of the USA) was constructed on an even larger area north of Piggots with an attendant base that included offices, houses, and amenities of a base to accommodate 350 permanent troops [67]. Antigua was able to supply most of the labour for the excavation work and construction of these facilities. There were also some available jobs on the base itself. Antigua, as a result, enjoyed a short burst of prosperity during the war.

Grenada, while not a "strategic island" like Trinidad, Antigua or St. Lucia, had its importance in the overall military strategy for the Caribbean. Pearls Airfield was a part of the chain of airfields that facilitated quick access to each island and provided emergency landing facilities for Allied aircraft. In Grenada, land for an airstrip was acquired at Pearls under the *Land Lease Agreement*. It was agreed that this airstrip would serve both military and civilian aircraft. In addition to the airfield, there was a

---

66   In time this complex would become Vieux Fort International Airport.

67   The entire area came to bear the name of Coolidge, and the airfield became the Vere Bird International Airport. The USA still has a lease on a section of the facility at Coolidge, and maintains a small base there.

navigational tower at the Pearls site with the call letters ZGT. The radio beam from this tower was stronger than from the towers in most other islands of the Lesser Antilles, because it was designed to be picked up as far away as Puerto Rico by military aircraft bound for Waller Field in Trinidad. Grenada was chosen as the site for this navigational aid, because the Northern Range in Trinidad blocked the radio beam from Waller Field. This navigational radio beam was also used by military aircraft flying to destinations in South America or West Africa. Point Salines Lighthouse was also an important navigational aid.

The first plane landed at Pearls Airfield in 1942, and the airstrip was formally declared open in 1943 for Grenadians. Having an airport and witnessing aircraft land and take off was a novelty, and families went on excursions to Pearls to see the airport and the planes. Bus tours were organised for people who did not have their own transportation. Margaret Phillip was taken on one of these "outings". Arnold Cruickshank and two of his friends wanted to see the airport and the activities there so badly that one day they rode their bicycles from St. George's to Pearls, and back home again. They chose the longer road through St. David's as it was less mountainous. This was a distance of about 17 miles each way. Arnold remembers being so stiff the next day that he could hardly walk!

Even where no American facilities were built, small parcels of land were leased to the Americans to be used as emergency landing strips. In St. Kitts, an emergency landing strip designed for aircraft unable to make it to Coolidge Airfield at Antigua was constructed and maintained in the middle of cane fields at Golden Rock. There were also American bases and airfields in Guyana, Curaçao and elsewhere.

There was an American Consulate in St. George's, situated upstairs in the building on the corner of the southern side of Young Street and the Carenage, where Rudolph's Restaurant used to be. The consulate maintained a powerful motor launch in a state of readiness to react to reports of U-boats near Grenada or in the Grenadines by setting off in search and pursuit. The launch was also used for visits to Trinidad by the Chargé d'affaires, Charles Whitaker [68]. Charles Whitaker's secretary

---

68 Charles Whitaker was elevated to the position of Vice Consul on 5th May, 1944 in recognition of his services.

was Nellie Donovan [69]. She remembers him as dapper and very nice. Once or twice she had the adventure of having to accompany Charles Whitaker on the nine-hour trip to Trinidad on this boat in the pursuance of official duties. Second in command at the consulate was John DaBreo, a Grenadian. He had first worked in the office of the American Base at Chaguaramas during its construction. When that job came to an end, simultaneously with the opening of the American Vice-consulate in Grenada, he was offered and accepted the position of Assistant to the American Consul.

Personnel of the U.S. military frequently visited Grenada, sometimes quite unannounced. On one occasion, a military vessel arrived with a number of U.S. military personnel, a couple of Mack trucks and heavy artillery. The vehicles were unloaded, and the artillery packed into them. The trucks were then driven roaring through the narrow streets of St. George's, ending their journey at Richmond Hill. The trucks and personnel in them returned a short time later, exactly as they had left, with nothing done with the artillery. Everything the military brought with them was re-loaded on the vessel, which then departed. The acting Colonial Secretary and Registrar General [70], Terrence B. Comissiong, knew nothing about the intended visit and, after lodging a complaint, was told that there was no need for concern, because the sites visited proved unsuitable for development for purposes envisioned!

There were two major gun emplacements in St. George's. One battery was at Ross's Point, and the other at Richmond Hill on the site of the former Lions' Den, now part of Her Majesty's Prisons. Soldiers of the Southern Caribbean Force manned these two batteries. The barracks of the Southern Caribbean Force at Tanteen and Ross's Point and the battery at Richmond Hill served to make the islanders believe they were protected. However, since the troops in Grenada were poorly equipped and had hardly any ammunition, Germany could have easily taken Grenada, or most of the "small islands" of the Lesser Antilles, if she had chosen to, but Germany was not anxious to be distracted at this time by the acquisition of colonies. Possession of the colonies would come later when Germany achieved its goal of domination of the world.

---

69   Mrs. Herbert Payne

70   Comissiong acted as Registrar General and Colonial Secretary from 2nd June, 1942 until 26th August, 1942.

During the war, small planes and sea planes of the British Fleet Air Arm in Trinidad would land on or near Queen's Park on a fairly regular basis to drop mail for the American consulate. The planes could be seen quite clearly from Moliniere, and were objects of interest for children such as Cecil Edwards and his playmates who watched them from the surrounding hills. They could also see the ships and tankers going to and from Trinidad and the sloops and schooners leaving St. George's harbour and going north. Cecil Edwards and his friends could identify most local boats by their shape and by the noise the engine made. Cecil remembers the distinctive shape of the *Island Queen* and the distinctive sound of its engine as it passed up the west coast on its frequent voyages to Carriacou and St. Vincent.

Other than the soldiers of the Southern Caribbean Force and the incidental incursions of military from Trinidad, military personnel were seldom seen in Grenada. However, occasionally both American and British warships would come into Grenada to visit and to give the crew much needed recreation. When a British warship visited, the Governor would entertain the officers at Government House. Young ladies from good families were invited to Government House to provide dancing partners for the officers. Officers were also entertained at parties at the Richmond Hill Tennis Club.

Crews from the occasional war ship would usually enjoy a friendly football match with a team of local young men, played on the Tanteen Field. Groups of school children were sometimes invited to tour visiting ships, but these visits were more usual after the war ended, when the war ships made their final visits to the islands.

The signs of war in Grenada were there in plain sight for all to see, but no evidence of war was as great as the evidence produced by the activity of the U-boats that entered the Caribbean silently and undetected in January 1942.

PART III

GERMAN U-BOATS IN THE CARIBBEAN

I: THE REWARDS OF CAREFUL PLANNING

From the first day of World War II, Grenada had experienced some of the exigencies and discomforts of being involved in a war. However, for Grenada, and the Caribbean, the first nine months of the war were ones of virtual quiet. This "phony war" lasted from September 1939 to May 1940. The only incident was in August 1940 when the *S.S. Davidson*, bringing bank notes to Grenada from Trinidad, disappeared, supposedly sunk by enemy action, although this was never proven[71].

On 7th December 1941 the Japanese destroyed most of the US Naval Fleet at Pearl Harbour and Manila in The Philippines. The very next day, 8th December, 1941, America officially entered Wold War II. Germany was now free to unleash *Operation Neuland* which had been carefully planned well in advance, and was absolutely ready for implementation. *Operation Neuland* was planned to inflict maximum disruption to allied

---

71 The loss was reported in the *Grenada Gazette* of 24th August 1940.

shipping in the Caribbean with attendant severe repercussions for the supplies. Oil, raw material for war purposes, food and other essentials were shipped from the Americas to Britain, and from one country to another within the hemisphere of the Americas. Manufactured goods, food, war material and other items were shipped from Britain and North America to countries within the region. The Caribbean was a network of busy shipping lanes. The U-boats that were to be the powerful weapons of this undersea war were ready to go and were now released to wreak havoc in an unsuspecting and unprepared Caribbean.

Each U-boat was commanded by a captain knowledgeable about the environment into which he was sailing. Adolf Hitler had been planning a global war of conquest since 1938 and had been thoroughly but surreptitiously preparing Germany's war machine. Germany had laid careful plans for a naval war in the Caribbean, and its strategy was already in a state of advanced preparation long before the beginning of the war. Unlike the Allies, who downplayed the strategic importance of the Caribbean, Germany fully appreciated that certain exercises were inevitable in this part of the world if a global war was to be won. The Southern Caribbean was the doorway to South America and also provided the approach to the many busy American ports along the Gulf of Mexico. Also, the tankers from Europe passed though the Caribbean, going to and from the refineries in Venezuela, the Dutch islands and Trinidad. Moreover, whichever nation or group of nations controlled the Caribbean also controlled access to shipping lanes in and out of the Panama Canal. The German objective was to disrupt marine traffic travelling through the Panama Canal between the Atlantic and Pacific Oceans, including warships going to and from the American and Asian theatres of war, thereby cutting off communication between Europe and the Americas. The German *Wehrmacht*[72] also sought to destroy the oil tankers and oil installations in the Caribbean, and to interrupt the trans-shipment of bauxite to the USA from Trinidad. Because of the shallow waters around the Guianas, shallow-draught craft brought huge quantities of bauxite down the rivers from the interior of British Guiana to Trinidad, where the bauxite was transferred to American freighters for delivery to the USA for the aircraft-building industries.

72　The *Wehrmacht* were the unified armed forces of Nazi Germany from 1935 to 1945. It consisted of the Heer (army), the Kriegsmarine (navy) and the Luftwaffe (air force).

U-boats had made their appearance during World War I, but as an important preparation for a new war, Germany had designed and built new models that were far superior and much more deadly. Another important part of Germany's advanced preparations for war had been to send naval training vessels to the Caribbean in the years immediately preceding the war. These would visit as many of the islands as possible. While in port, the cadets on these vessels were instructed to investigate the harbours thoroughly, and learn the general geography of each island. The soundings and measurements they took, and the numerous photographs taken, were all thought by onlookers to be part of the training exercises, or the normal behaviour of tourists. These cadets, however, were destined to be the captains of the U-boats.

Germans had also sailed the Caribbean in command of ships of the Hamburg-America Line, which were frequent visitors to the islands of the Caribbean and the East Coast ports of the United States. When the war began, therefore, Germany not only had up-to-date maps provided by their merchant fleet, but they also had the men who knew firsthand and intimately the sea conditions in the Caribbean — the locations of reefs and rocks, the details of the harbours, lights and currents, and preferred routes of merchant ships. The younger commanders of the U-boats to be sent to the Caribbean were thoroughly briefed by the merchant captains and others who had this wealth of Caribbean experience.

II: THE DEFENCELESS CARIBBEAN

The threat that Hitler's ambitions for Nazi Germany presented to world peace was evident to Britain and her Allies, but the British Government hoped that they could avert a war through diplomacy. In the years preceding the outbreak of World War II, Neville Chamberlain, the British Prime Minister, followed an appeasement foreign policy, adamant that Adolf Hitler could be appeased and war averted. In 1938 he signed the *Munich Agreement* conceding the *Sudetenland* region of Czechoslovakia to Nazi Germany in the hope that Hitler's expansionism would stop there. Hitler, however, was bent on a world war of conquest that would end in Nazi domination of the whole world. There was no stopping him with the sacrifice in Czechoslovakia.

Chamberlain and his Government were also against overt preparations for war, because they felt that preparations for war would jeopardise Anglo-Germanic relations. Chamberlain's policies were very popular in England before the war, but with the advantage of hindsight, Chamberlain is today generally criticised for his pre-war policy and the resulting lack of preparedness for World War II. Bousquet and Douglas write:

> A year before the declaration of war, in September 1938, Chamberlain was just back from Munich having sold Czechoslovakian freedom for the price of transitory peace. Confident of maintaining peace, it was not until the summer of that year that he began to prepare seriously for war. But the prime minister's reluctance to appear aggressive towards Germany meant that Britain's war planning, in the years preceding 1939, consisted of half-baked preparations carried out in a half-hearted manner. [73]

When Hitler invaded Poland, Chamberlain realised his mistake in believing that Hitler's expansionist plans had been limited, and he was forced to declare war, keeping a promise to defend Poland. Eight months later, Chamberlain and his Government faced a vote of no confidence in Parliament, and he was forced to resign. On 10th May, 1940, Chamberlain's conservative Government was replaced by a coalition Government headed by Winston Churchill.

However sympathetic one might be to Chamberlain, who wanted to spare England another costly and bloody war, the consequences of delay were that Britain and France were caught with their programme of building destroyers and other battleships, including those to be used as escort vessels, still in progress. Moreover, the Allies believed that they would be able to concentrate on activities on the European front and to provide escort for ships carrying material and men across the Atlantic to Europe. They believed that the Americas and the Caribbean were safe from German attack, estimating correctly that the German *Luftwaffe* did not have the capability to reach targets in the Caribbean and return to Germany. Surveillance on South American countries was constant,

---

73   Bousquet and Douglas, P. 17

especially with respect to those friendly to Germany, to guarantee that they did not provide landing and refuelling facilities for German planes. The Panama Canal was also defended — otherwise no necessity was seen to protect and defend the Caribbean. Although the Allies had planned to convoy supplies and troops across the Atlantic, these plans did not provide for the protection of Allied shipping travelling through the Caribbean.

The Allies were also completely unaware that Germany had the capability to deploy submarines in the Caribbean, and took no anti-submarine measures in this region. This mistake would cost them dearly.

In contrast, Germany had planned meticulously for a war that encompassed the entire globe, in all possible theatres of war including the Caribbean. When the *Operation Neuland* began with the sinking of thousands of tons of American and Allied shipping in the Caribbean, and Germany's intention became plainly evident, it was almost too late for the Allies to recover the situation. The plan to use U-boats in the Caribbean was totally unsuspected by the Allies. The U-boats in the Caribbean did inestimable damage to the Allied war effort, demonstrating that Germany had discovered that the Caribbean was the Allies' "Achilles Heel".

Submarines had been used by both Germany and Britain in World War I. The allies knew that improved models of German submarines[74] were under construction since 1935. They also knew that the German U-boats did not have the range to reach the Caribbean and return to Europe. Where Allied intelligence was lacking was that Germany was building submarine tankers that were totally devoted to refuelling the operational U-boats. Nicknamed *Milch Cows*, each of the XIVC U-cruisers was designed to carry 700 tons of fuel, which was enough to refuel 12 medium-size U-boats or five of the more advanced type IXC U-boats, the latter of which was the type of submarine that usually operated in the Caribbean. The *Milch Cows* also carried a stock of essential medical and other supplies to replenish the stocks of the attack vessels in cases of emergency. Most U-boats of the type IXC had a fuel capacity range that allowed them to remain on station in the Caribbean

---

74 Specifications for British and German submarines and other vessels are to be found in many places. For a concise account of the specifications and establishment for the Type IXC U-Boat which was the most common type used in the Caribbean, see *Metzgen and Graham*, P. 170-171.

for 3 weeks, not including the week's journey to and from their base at Lorient. The capacity to refuel at sea therefore gave the U-boats access to the furthest reaches of the Caribbean and the Gulf of Mexico for extended periods of time. Believing that the U-boats did not have the range to reach the Americas and the Caribbean, the Allies left the Caribbean completely defenceless at the start of the offensive.

Another gap in Allied intelligence was the speed at which the U-boats could travel. The *corvette*, which travelled at 15 knots and which was designed as an anti-submarine vessel, was too slow to be effective against U-boats, which travelled at 18½ knots. Even the boats that Britain had newly built were wrong for the service that they now were pressed to perform.

The IXC submarine had a complement of 48–57 officers and men, and 7 torpedo tubes — 4 in the bow and 3 in the stern. It carried a stock of 24 22-inch torpedoes each with a 600 pound warhead. Each U-boat was further armed with one 37 mm gun and two twin 20 mm anti-aircraft guns. U-boats could crash dive in 35 seconds. The hulls of the U-boats were built of Krupps steel, and could withstand water pressure at 600 feet. Some could survive at even greater depths. To damage a submarine fatally, a depth charge had to detonate within 30 feet from the hull, and bombs dropped from the air had to detonate within 6 feet of the hull. The U-boats had diesel engines that gave them a top speed of 18½ knots on the surface, and electric engines that allowed them to travel underwater at seven knots – slightly faster than the speed of the average convoy.

At the beginning of the war, the German and Italian submarines were basically surface vessels that had the ability to submerge, and remain under water for an amount of time, limited by the air supply in the boat and the capacity of the batteries. The tactic of most U-boats was to stay on the bottom of the ocean in daylight in relatively deep water, surfacing and moving in shallower coastal waters during the night, when they would recharge their generators and replenish their air. Even while at war, Germany assiduously worked to improve the submarines and also re-fitted older submarines with new devices to improve the efficiency of the crafts and to defend against the improved submarine detection technology of the Allies. At the very end of the war, Germany unleashed her new ultra-modern submarine, the type XXI. By then, however, Germany had completed its operations in the Caribbean.

By the beginning of 1942, Germany had over twenty U-boat bases, most of them located in occupied countries. Nearly all of the U-boats operating in the Caribbean were based in Lorient, France, the second-busiest U-boat base, the first being that of Kiel in Germany. Each U-boat was known by its number, and some of these numbers would never be forgotten for the damage they did and the fear they spread once the U-boats were unleashed against Allied shipping in the Caribbean.

The commander of the U-boats was Grossadmiral Karl Dönitz. [75] The plan was that through the deployment of the U-boats, no supplies from the Caribbean would to get to England and none to come to the Caribbean, and Britain would be brought to her knees. Under his direction, *Operation Neuland* began spectacularly. The crews of the U-boats were ordered to concentrate on sea traffic moving around the Dutch islands of Curaçao and Aruba, with their large oil refineries, and Trinidad, which produced its own oil. They were also to cover all the shipping lanes and sink as many tankers and merchant ships as possible belonging to the Allies that they caught moving in or out of the Caribbean.

During the night of Monday 16th February, 1942, there was a successful attack by U156 commanded by *Korvettenkapitän* Werner Hartenstein [76] on oil tankers at Aruba. The *S.S. Pedernales*, which was anchored about quarter of a mile offshore, was also hit and burst into flames. Minutes after, the *S.S. Orenjestas*, which was anchored nearby, was hit and sunk. Three more tankers were sunk: two British and one Venezuelan, and two tankers were damaged: one Dutch and one American. U156 also shelled the storage tanks on shore, but the shells could not penetrate the tanks. However, damage was done to the compound.

---

75 Karl Dönitz (16th September 1891 – 24th December 1980) served in the Imperial German Navy during World War I. During World War II he commanded first the German submarine fleet, and then the entire German Navy. In the final days of the war, Dönitz was named by Adolf Hitler as his successor, and after the Fuehrer committed suicide, the admiral assumed the office of President of Germany. He held this position for about 20 days, until the final surrender to the Allies. After the war, Dönitz was convicted of war crimes at the Nuremberg Trials and served ten years in prison.

76 Werner Hartenstein, a veteran of the First World War, did immense damage to Allied shipping during the 294 days he operated in the Caribbean. His totals were 20 ships sunk and 4 ships damaged, including a warship. Hartenstein and the entire crew of U156 were killed in action on 8th March, 1943, when struck by depth charges from a US PBY Catalina aircraft in the Atlantic, east of Barbados. He was 35. He was highly decorated during his service, the highest award being the Knight's Cross in 1942.

Another U-boat was in place to let go torpedoes on ships in the harbours at Curaçao and Maracaibo, and to shell the oil installations on the same Monday night. Two nights later, two ships were torpedoed in the Gulf of Paria. Robert Ferguson, a Grenadian working as a clerk in the technical department of the Largo Oil refinery in Aruba, witnessed the shelling of the Largo Oil compound. The office buildings were left in ruins, but fortunately the shelling killed no one. He also recalled that supply ships bringing food to the compound were torpedoed on several occasions.

In the nineteen days of January 1942, thirty-nine ships were sunk by U-boats in the Caribbean, sixteen of them tankers. During 1942 and the first half of 1943, the U-boats were relentless in their attacks on shipping along the coasts of North and South America and in the Caribbean. The worst damage was done in 1942. The Allies and neutral countries lost approximately 350,000 gross tons of merchant shipping in the Caribbean and Gulf of Mexico in the two days between 16th February and 17th February, 1942. From 16th February to 31st December, 1942, the U-boats sank 337 ships in the Caribbean, the majority being American. This averaged out at 1.45 ships per day.

For the first few months of U-boat activity in the Caribbean, the Allies did not know how many U-boats had been deployed or where they were. Shipping did not know the elementary ways in which they could have protected themselves. Thus, Allied ships travelling along the North American coast or in the Caribbean Sea and the Gulf of Mexico travelled unescorted and displayed full markings, navigational and other lights. This made them "sitting ducks" for the U-boats. Kelshall observes that:

> The U-boat commanders had difficulty believing that they were operating off the coasts of a nation at war. Independently sailing merchant ships, silhouetted against a brilliantly lit coastline proved to be ideal targets for the U-boats. [77]

Bauxite shuttles, the freighters in which the bauxite was transferred for transport to the USA, oil tankers, ocean freighters bringing military equipment to Trinidad and the region, vessels carrying sugar or other raw materials to Britain, boats from South America carrying sugar, coffee and other products for consumers in the United States, vessels bringing food and other supplies from Europe to the Caribbean, and the

---

77   Kelshall P. 13

occasional passenger liner or troop carrier all were sent to the bottom of the sea by the U-boat action, and the Caribbean and Atlantic quickly became a graveyard for Allied shipping. When the Allies changed the shipping routes as an avoiding action, the new routes were quickly discovered by Germany, and additional U-boats dispatched to cover the new areas.

Kelshall identifies three "chokepoints" – areas in the Caribbean where ships would congregate for one reason or another. These would be the areas where the U-boats would lie in wait to reap a grim harvest. The Windward Passage, a strait between the easternmost region of Cuba and the northwest of Haiti, was a chokepoint, because ships had to pass through this narrow piece of sea between Cuba, Jamaica and Haiti to get in or out of the Northern Caribbean. The second chokepoint was the area around Aruba where ships would collect oil from the refineries in Aruba and Curaçao, and then head to Trinidad to join the convoy. Trinidad and the waters around it as far north and east as Barbados and as far west as Grenada constituted the third chokepoint. An area of this third chokepoint which formed a right angled triangle with the Dragon's Mouth at one angle, the Serpent's Mouth at another, and Grenada as the third angle with Tobago encompassed, was the Southern Caribbean's own Torpedo Junction.

Grenadian historian Fitzroy Baptiste, an authority on the Second World War, writes as follows:

> In 1942-1943 a 150 mile strip around Trinidad suffered the greatest concentration of shipping losses experienced anywhere during World War II. During 1942 some 43.3 ships were sunk for every U-boat in the circum-Trinidad area, but only 9.2 world-wide. The total number of ships exclusive of schooners and other small craft ... sunk by the U-boats in the Trinidad arc by the end of 1942 was 130 or 750,000 tons of shipping. Approximately three-quarters of those ships were cargo vessels. They included bauxite shuttles from The Guianas to Trinidad. Just less than one quarter were tankers.[78]

Because of the total lack of defence against submarine attack on shipping, German U-boats "played" at will, unmolested for almost two years until adequate defences could be devised against the U-boats.

---

78 Baptiste 1988, P. 144

The shipyards of America and Britain would try to build new ships fast enough to replace the ones lost. Early in 1942, US President Roosevelt inaugurated the largest ship-building programme in history, building an average of three ships a day, but there was a deficiency gap that just could not be closed. In May and June 1942, almost double the amount of ships completed was sunk in the Caribbean.

III: ACHILLES' ESCAPADES

Korvettenkapitän Albrecht Achilles [79] was the young German captain of U-boat 161. Now 28 years old, he had visited many of the Caribbean harbours as a naval cadet on a Germany training ship, and thus knew some of the harbours very well, including those of Trinidad and St. Lucia. He would put this knowledge to good use in two really daring escapades in the Caribbean.

On 16th February, 1942, Achilles in U161 made an extremely bold entry into the Port-of-Spain harbour. Slipping through the anti-submarine magnetic detector loops stretched between the islands of the Bocas, he sailed undetected into the harbour. Minutes before midnight, torpedoes from his U-boat found home in the side of the 7,400 ton American freighter *Mokihana* and the British tanker *British Consul*. Both ships settled into the mud, with holes torn in their sides and the *British Consul* on fire. Kelshall[80] describes the twin blasts rolling like thunder across Port-of-Spain, and Achilles exciting escape.

Aware of the slim chances of escape from a harbour now in uproar, Achilles dived as soon as he was sure that his torpedoes had found their mark. To his dismay, the bow of the submarine stuck in the mud. Successful in using his engines to free the submarine, he realised that the water was too shallow for a submerged escape. He had to make a

---

79 This daredevil submarine captain and the entire crew of U161 lost their lives when the U161 was sunk off the east coast of Brazil on 27th September, 1943 by a PMB Mariner aircraft out of Trinidad flown by Lieutenant Harry B. Patterson. Achilles was 29.

80 Kelshall P. 40-41

run for it on the surface. Sheltering in Diego Island, Achilles waited for the procession of launches he observed running east to west along the south shore of the naval base at Chaguaramas. He fixed his speed to that of the launches, and ran with just his conning tower above the water, blending into the parade. He crossed Chaguaramas Bay just as if he was one of the small craft belonging to the United States or Royal Navy excitedly moving about at this time, his running lights resembling those of a small launch. He took U161 close to shore underneath the batteries on Gaspar Grenade, passing within a hundred feet of the shore. If anyone saw U161, they thought she was a navy launch, legitimately there. U161 continued her journey, running the hurdle of the fortifications at Monos and Chacachacare, and Huevos. Forty-nine minutes after the torpedoes struck, Staubles Bay detected a submarine passing over the anti-submarine magnetic detector loops. Achilles was departing. U161 submerged, but not before seeing searchlights sweeping the waters ahead looking for her. Once out of the Gulf of Paria, Achilles sailed at full speed, free in the Caribbean, leaving behind every anti-submarine vessel and B18 bomber searching the Gulf but not outside the Bocas for U161. By the time a massive search was instituted outside the Gulf, U161 had long gone. [81]

On another occasion, an unidentified U-boat dared to venture into the heavily protected harbour at Point Fortin. John Watts remembers the terror that the sight evoked in the population, for should the U-boat shell the installations of the United British Oilfields in that port, there would have been massive explosions and loss of life. Luckily for John Watts and the others who lived in Point Fortin at the time, the submarine was chased away by armed launches and aircraft from the American bases before it could do the sort of damage that had been inflicted on the Port-of-Spain docks.

Three months later, Albrecht Achilles took his U-boat right into the harbour in Castries, and sank the *Lady Nelson* and the *Umtata*. At 11.35 p.m. on the evening of 9th March, 1942, the lookout at the Vigie lighthouse, Police Constable Joseph Rachael spotted what he took to be a small craft, maybe a schooner, moving into the harbour directly below him, but a look through his night binoculars revealed the long dark shape of a submarine slinking silently into the harbour with its

---

81 Bertrand Pitt informed the author that for all his life he believed what he had been told at the time – that U161 has not escaped from Port-of-Spain harbour, but had been "caught in a net"!

conning tower above the water. It was later identified as U161 under the command of Albrecht Achilles. Rachael hastily called his headquarters at the dockside, but it was too late. At 11.45 p.m., heavy explosions rocked Castries, echoing from the surrounding hills.

The *Umtata* was a coal-burning merchant steamship of the British Bullard-King Line. It had put into St. Lucia for a supply of coal, and then it would have joined a convoy for the United Kingdom. The *Lady Nelson*, one of five identical Lady Boats in Canadian registry, was bound for Barbados. Both were at the north dock and both were hit by one torpedo each, first the *Lady Nelson* and then the *Umtata*. They burst into flames, with holes ripped into their sides, but the harbour was too shallow to allow them to sink, so instead they settled into mud, fortunately remaining upright.

Fifteen passengers and three crew members were killed on the *Lady Nelson*, most of them "deckers" who had boarded in St. Lucia and were travelling to Barbados. Four sailors from the engine room were killed on the *Umtata*. [82] Seven dockworkers also perished. The rolling thunder of the explosions shattered all the glass windows in Castries and damaged houses and buildings on the waterfront. The precincts of Castries harbour was a scene of pandemonium as soldiers and citizens alike rushed to help to extinguish the raging fires on the affected ships and to attend to the injured and dead.

Paul Slinger, at the time a child of seven or eight, remembers very well the night the *Lady Nelson* and the *Umtata* were torpedoed. His father was a medical doctor in St. Lucia stationed at the hospital near Castries. The doctor's house was in the hospital compound, which overlooked the sea. Paul was fast asleep when a gigantic explosion threw him out of bed to the floor. Discovering what had happened, his father rushed to report for duty immediately and did not come home for days. Carol Bristol also remembers that night and the terrified uproar in the population when the boat was torpedoed.

As to U161 and its crew, Captain Armstrong notes that:

> His powerful Diesels could clearly be heard by residents of Vigie immediately after the explosions and those who rushed to places where the Harbour could be seen were rewarded

---

82   Some accounts give the causality figures for the *Umtata* as eighteen.

with a quick glimpse of the U-boat turning on the surface on its way to the open sea.[83]

As the U-boat passed through the narrow entrance to the Castries harbour, machine-gunners of the Southern Defence Force at Meadows Battery and Tapion Rock fired at it with the ammunition they had, but these were not armour-piercing shells, and the unsuitable bullets simply bounced off the hull. One of these gunners is said to have been Ben Jones [84] from Grenville, Grenada.

It is said that a great shout went up from the crew of U161 when word was sent down that they had successfully completed another daredevil attack. Achilles had been prepared to make a strategic sacrifice in coming into a harbour that was so tightly enclosed, but the sacrifice was more of a calculated risk, which paid off. Achilles knew Castries harbour and all its characteristics well, for he had visited St. Lucia in 1936 on board the German training ship *Schleswig Holstein* and, under the cover of enjoying what St. Lucia had to offer, had extensively photographed the harbour, both from the perimeter and from the Morne and la Toc and Vigie peninsulas. The cadets from another training vessel, the *Gorch Fock*, on their visit to St. Lucia in 1938 had taken soundings of the harbour, and up the west coast. Achilles was also aided by the lights of Castries, which were ablaze in spite of the war. These had illuminated his two targets. The authorities in Castries had been complacent and naïve and this important harbour was poorly defended. But however easier the task was made, Achilles' sortie into Castries harbour was extremely daring, and the saga of the sinking of the *Lady Nelson* and *Umtata* lives on when most war stories from the Caribbean have been forgotten. One of the stories told is that Achilles is supposed to have said that he could have shelled and destroyed the whole of Castries if he had wanted, except that he had been entertained as a cadet officer from a training ship by gracious St. Lucians, and was returning a favour in view of their hospitality. Blackouts were instituted after this incident for the entire Caribbean to prevent the lights of the town and the lights of the ships ever again being beacons leading the U-Boats to their targets.

The *Lady Nelson* and *Umtata* blocked the harbour and access by other boats to the pier, and until they could be repaired and refloated, cargo had to be offloaded in the outer harbour into small boats and transferred

---

83 Armstrong, P. 39

84 Ben Jones became a popular lawyer and politician. He served as an interim Prime Minister of Grenada.

in "dribs and drabs" to the pier. From the night of the 9th March onwards, the population of St. Lucia, and in fact the entire Eastern Caribbean, never again felt that the war was a far-off event. Everyone was more alert, and as a precaution against this sort of attack even re-occurring, an underwater net was strung across the mouth of Castries harbour.

The people of the Eastern Caribbean took the attack on the *Lady Nelson* as a personal affront. Lady Boats were much beloved by people in the islands and valued for the transport they provided between the islands as well as for their cargo service. Operated by the Canadian National Steamship Service since 1928, they were multi-purpose vessels, built to carry cargo between Canada and the Caribbean, to accommodate tourists in luxury, and to provide cheap "deck" passages between the islands to West Indians. The destinations of these boats included their home base, Canada, and Boston, Bermuda, St. Kitts, St. Lucia, Barbados, Trinidad, Grenada and Georgetown, British Guiana.

The *Lady Nelson* used to visit Grenada once a week, and was a part of Grenadian life and psyche. Children were welcomed visitors on the Lady Boats. Geraldine Sobers remembers that they were given red apples when they visited the boat. To hit a Lady Boat was really to hit home. One respondent admits that as a young girl she didn't "take the war on" until she heard that the *Lady Nelson* had been hit. Soon a calypso [85] was composed commemorating the sinking of the *Lady Nelson*. The words are remembered to this day by Carlyle John and other people who were children at the time:

> Lord bless *Lady Nelson*
> A submarine a-chase her
> Right in the harbour.
> As soon as she anchor
> He bomb her *pou désann o fon*. [86]

Which Lady Boat was it that students at the Church of England High School for Girls in Grenada remember rushing into the St. George's harbour running from a U-boat, sometime in 1942 seemingly making straight for buildings on shore? Was it the *Lady Nelson* before she was sunk? Whichever *Lady* it was, it stirred up a lot of mud in the St.

---

85   Other calypsos were composed and sung by Grenadians during the war, many being the compositions of the famous calypsonians of Trinidad. Among the other calypsos remembered was one called *Mussolini and Hitler*, and other called *What Chamberlain Said*.

86   Patois meaning "to sink to the bottom".

George's harbour, but thankfully, did not get herself stuck. She stayed in the harbour just long enough to make sure it was safe for her to proceed on her journey.

The *Lady Hawkins* was the first Lady Boat to be lost in January 1942, torpedoed off the coast of Bermuda in the middle of the night, with only 75 survivors. On 9th March, 1942, three months after the *Lady Hawkins* was torpedoed, the *Lady Nelson* [87] was torpedoed in the Castries harbour. Two years later, on 8th May, 1944, the *Lady Drake* was torpedoed between Bermuda and Halifax. Twelve persons were killed, and the 260 survivors were later picked up from their lifeboats. This boat was carrying fifty Barbadian artisans selected for employment in Britain's war industries. Luckily, none were among the unfortunate twelve. By the end of the 1944, Grenada and the rest of the Caribbean got the very sad news that Canada was forced to withdraw all the Lady Boats from service, as three of the five had been torpedoed. The service to the Caribbean would be resumed after the war.

IV: SIRENS AND BLACKOUTS

After the *Lady Nelson* was torpedoed, blackouts were instituted and strictly policed not only in the towns, but on the ships and boats in the harbour, and in the town's suburbs all over the Caribbean. Whether there was time for a warning or not, Grenada would be prepared for any eventuality. In Grenada all lights were to be put out by 9 p.m, but the population was discouraged from using lights at all in their homes after dark. Thick, dark curtains were hung at the windows before

---

87 The *Lady Nelson* was raised and towed to Mobile, Alabama. She was re-launched in April 1943 converted into a hospital ship, her cabins removed to make room for hospital wards and medical laboratories. The *Lady Rodney*, the last remaining liner of the Canadian *Lady* Boats, was taken over by the Canadian Department of National Defence in June, to serve as a troopship between Canada, Newfoundland and Labrador. At the end of the war, both these ships were used to repatriate Canadian soldiers and their families from Europe. They were finally reconverted into passenger liners for Caribbean cruises. In July 1947, the *Lady Rodney* sailed from Halifax to the Caribbean, and the *Lady Nelson* followed a month later. In March 1952, it was announced that these two *Lady* Boats would be retired at the end of the summer season of that year, because they could no longer compete with more modern vessels.

any lights were lit. In addition, sometimes the window panes were painted black or dark green to further hamper the escape of any light from the house.

Brenda Wells relates how careful the family was with lights as their house in La Borie faced Calivigny harbour. Government House was not exempted from these precautions. Adina Maitland, who was taken as a very young girl to assist Lady Grimble, the Governor's wife, remembers that the windows of Government House were shuttered and hung with dark curtains. Kerosene lamps were used instead of electric lights. She remembers that there were many blackouts.

Outside of town and in the country few had electricity, and used "pitch oil" (kerosene) lanterns, such as the Coleman lamp, the Tilley lamp and the Aladdin lamp. The bigger children, particularly the boys, were given the duty of cleaning the chimneys and trimming the lamp wicks every day, unless there was a servant to do this. Arthur Bain, who lived at *The Willows* in Grand Bras, was familiar with these lamps, and remembers that the Coleman and Tilley lamps had to be pumped. Pumping the lamps was the job of the biggest children or an adult. The lamps were not safe for children to light, and this was always done by an adult.

The necessity to use as little light as possible applied equally to those with lamps and those with electricity. The chimneys for the lamps in Leo Cromwell's home in Woburn were partially painted red to block most of the light the lamps gave. Reginald Dowe lived on Richmond Hill with his grandmother, who kept the lamp light low and, in addition, placed the lamp in a big biscuit tin to further stifle the light that could be seen. Candles were used in some homes, but in very poor households, no lights were used at all. People went to bed at sundown, and rose at dawn.

There were only a few dim street lights in the towns, and these were also covered with kerosene tins so the light from them was dimmed and directed straight down. The population was asked to travel at night only in emergencies. When they did, owners of vehicles were required to partially cover their headlights, or to direct the beams downwards and to drive on dim.

Blackout drills were also scheduled regularly in St. George's after 1942. On occasion, there was an unscheduled blackout that was very frightening. This would sometimes happen when a U-boat was sighted in the harbour, and looked like it might "mean business". The general population knew what was required of them. For both the scheduled drills and emergency blackouts, a siren would sound and could be heard for miles. As soon as this went off, all street lights, domestic electric and other lights within a mile radius of St. George's had to be extinguished. Any vehicle caught on the road during a full blackout drill was required to douse its lights, park at the side of the road, and wait for the blackout to end. Pedestrians on the street would seek refuge inside the nearest building. People were advised to respond as if they were going to be attacked from the sea or air every time they heard the siren, because there would be no way of warning them of a real event. The sirens also sounded at the end of the blackout period.

The children hated the blackouts. Some parents tried to ease the trauma of the blackouts by inventing a pleasant ritual to distract the children. Most families would tell stories until the blackout ended. Jenny Yearwood, whose family lived at the Mardi Gras Junction in St. Paul's, remembers that her mother had a special tin of sugar cakes, ground nut cakes and ground nuts, which would appear whenever there was to be a blackout. These were passed around while stories and jokes were told. Monica Knight, a nurse and a neighbour, would spend the night and join in the fun whenever there was a blackout.

Among the duties of the Grenada Volunteer Force and Reserve was to serve as blackout wardens. The fathers of Judith Parke, Marcella Lashley[88] and Brenda Wells were among the volunteers in the Reserve guard who went to regular practice drills in the early evening, and served on occasion as blackout wardens. The reservists wore uniforms when they were on duty. Ruby DeDier and Margaret Phillip both remember the heavy steps of the wardens pacing the streets to ensure that no lights would guide the German U-boats into the harbour of St. George's. The wardens would knock at the doors of anyone who had even a chink of light showing. A respondent remembers a warden knocking at her mother's door on Tyrrel Street [89], saying that he could see a light. Her mother, a very proud woman, got in a temper, and demanded that the warden come and show her from whence the light was escaping! Sometimes the reservists patrolled by van in the rural areas when there was a lot of ground to cover — too much to do by foot patrol.

---

88 Mrs. Chasely David

89 In the 1990s, Tyrrel Street was renamed H.A. Blaize Street to honour Herbert A. Blaize, one of Grenada's Prime Ministers.

When the sirens blared for scheduled or unscheduled blackouts, all reservists who lived close enough to hear them had to drop whatever they were doing, leaving their regular jobs if necessary, and report to the Drill Yard *post haste*. On occasion, the sirens blew during the day. Regardless, all reservists and the army cadet corps had to report for duty immediately.

V: ATTACK IN BRIDGETOWN HARBOUR

Another spectacular attack in a Caribbean harbour took place on Friday, 11th September, 1942, between 4.30 p.m. and 5.45 p.m. when the freighters *Cornwallis* and the *Betancuria* came under torpedo attack in Carlisle Bay, Bridgetown, Barbados by U-boat 514 under the command of Kapitänleutnant Hans Auffermann [90]. The strikes were not easy to achieve. Protective anti-torpedo nets had been provided for Trinidad and Barbados after Achilles's attacks on Trinidad and St. Lucia. These first had to be breached.

The first hole in the submarine net was blasted with four 600 pound warheads. Great fountains of water rose upwards, together with the remains of the net and the buoys. The tremendous detonation brought hordes of onlookers to the shores of the bay to watch. They were in time to see the *Betancuria* fire the guns, which, like many merchant vessels, she had been fitted during the war for protection. The guns were fired in the general direction of U514, but the submarine stayed submerged and so was safe from gunfire. Four torpedoes were fired at the *Betancuria*, but all four missed their target. The U-boat then blasted a second hole in the net to enable an attack on the Canadian freighter *Cornwallis*. The considerable amount of debris, including many of the buoys supporting the net, which was blown sky high by the explosion, must have clearly made it apparent to the attacker, and indeed all observers, that the net

---

90 Auffermann had visited Barbados just before the war on the training ship *Scheswig Holstein*. This bold U-boat captain was only 29 when he was killed with all his crew on 8th July, 1943, when his boat was sunk by rockets from a British Liberator aircraft north-east of Cape Finisterre, Spain.

had collapsed, or was otherwise damaged sufficiently to enable another torpedo to reach its target.

For some reason, *Cornwallis* did not fire her guns until after the net was breached and she had been hit. This vessel sustained serious damage. A gash 44 feet long and 14 feet deep was made in her side in the region of her No. 3 hold. The adjacent engine room bulkhead was shattered, and the engine room damaged. The sea rushed in, flooding the boiler room and several of her compartments. As the *Cornwallis* settled into the sand, cargo washed out of the great gash in her side. Luckily, although some seamen were injured, none were killed. As Auffermann departed with his submarine unscathed, he left considerable state of alarm and confusion behind in Barbados.

The cargo that has washed out of the *Cornwallis* was later dived up by intrepid Barbadian swimmers. The *Cornwallis* was repaired, only to be torpedoed again and sunk near Maine, USA, two years later.

VI: OTHER SHIPS SUNK NEAR GRENADA

The attacks on the harbours of Port-of-Spain, Castries and Bridgetown were "close to home" for Grenada, for not only were they geographically close, but Grenadians had close social and familial ties with the people of these islands. Many other ships were torpedoed within a 90-mile radius of Grenada, and people living on the hills of the parishes of St. George and in St. David will never forget the horror of seeing ships burning on the horizon from their homes.

In the months of April and May 1942, the German U-boats claimed many targets around Grenada. In April the *Korthian*, a small freighter, was sunk just south-east of St. Vincent on the 14th, the *Amsterdam*, a 7,400 hundred ton freighter was sunk 60 miles south-west of Grenada on the 16th, and the *Harry G. Seidal*, a large tanker, was sunk 80 miles west of Grenada on the 29th. Then in the month of May, the *Sandar*, a Norwegian tanker, was sunk east of Grenada on the 2nd, the *Brabant*, a Belgian freighter of 8,200 tons was sunk south-south west of Grenada.

On the 14th, the *San Victorio*, a large British tanker of 8,000 tons, was sunk south-south west of Grenada and 80 miles from Trinidad on the 16th. The next day, 17th May, the *Challenger*, an American freighter went to the bottom 15 miles north-east of Grenada, and then the *Sylvan Arrow*, a large Panamanian tanker of 7,000 tons, was sunk about 60 miles south-south east of Grenada on the 19th.

On 22nd May, 1942, the freighter *Watsonville* sailed out of Kingstown harbour with a cargo of St. Vincent arrowroot and cotton. Three miles from shore, it was torpedoed by U155 under the command of Korvettenkapitän Adolf Piening. [91] Edgar Adams reports that the final minutes were viewed by a crowd of terrified onlookers who had quickly assembled at the waterfront at the sound of the explosion. They were in time to witness the amazing sight of the sinking ship [92].

The devastation of shipping continued. The 7,000 ton freighter *Arkansan* sank west of Grenada on 15th June, 1942, and on the same day *West Hardaway*, a freighter of 6,000 tons, was sunk south of Grenada.

U-boat Captains usually let local Caribbean shipping pass unmolested, since their directive was to concentrate on sinking huge tankers and cargo ships of considerable tonnage. Local boats were of little consequence in terms of tonnage sunk. But there were exceptions. An expensive torpedo fired at a wooden schooner would be overkill — the schooners were easily sunk by shells fired from the U-boat's armament. One shell from a submarine's 20 millimetre gun was sufficient to put a sizable hole into the side of a wooden hull, and that was usually the end of that boat. U-boats were generally on the surface when they challenged a schooner, and at least 4 submariners were always on the conning tower while the U-boat was on the surface keeping watch. When a schooner in which the Germans had interest was spotted, the U-boat might approach, or wait until the schooner came within hailing distance. At least one crewman on each of the U-boats in the Caribbean spoke perfect English. If the intention was to sink the vessel, crew on the U-boat would instruct the crew and any passengers on the schooner through a hailer, that they would be given a few minutes to get into their lifeboat before they were shelled.

---

91   Korvettenkapitän Adolf Peining survived the war. He died in 1984 at the age of 73.
92   Adams, 2002 Pg 168

On Sunday, 28th June, 1942, there was a fateful encounter between U126 captained by Korvettenkapitän Ernst Bauer [93] and the schooner *Mona Marie*, owned and captained by Laurie Hassell of Barbados. The *Mona Marie* was a 142 foot two-masted schooner built in Nova Scotia. She plied the Barbados to Trinidad route and was about 55 miles south of Barbados, and north-east of Grenada, bound for Trinidad with a cargo of empty gasoline drums, when she was spotted by U126 on the surface, with crew on the conning tower, heading directly for her. The crew hailed the *Mona Marie*, but some movement aboard the schooner must have been misinterpreted as hostile, causing the U-boat to fire. The masts were hit, the sails torn, and two men were injured by wood splinters. The lifeboat was also punctured.

After some dialogue between Hassell and the crew of the U-boat, during which Hassell was told to cut down the remaining sails and to hand over any food that was on board, Hassell, his crew and his passenger were allowed to get away on the 14-foot dingy while the U-boat shelled the *Mona Marie*. However, she would not sink, as she was buoyed up by the empty gasoline drums!

Rowing west nonstop, the overcrowded dingy from the *Mona Marie* reached Mustique after traversing 110 nautical miles in 31 hours. Although they had been able to take some rations with them, the five men arrived sun burnt, dehydrated, hungry, and completely exhausted, but they were alive.

For a short while, the *Mona Marie* was the Caribbean's "Ghost Ship". She drifted into the Grenadines, but nobody would board her, as they regarded her as spooked. She was eventually sunk by a naval vessel from Trinidad, because she was a hazard to navigation.

Another schooner known to have been sunk by a U-boat was the *Seagull D*, a 76-foot two-masted schooner from St. Vincent, recently built by Randolph Adams and owned by S.G. "Papa" de Freitas. On 18th August, 1942, the *Seagull D* left the St. Vincent harbour with a full load of passengers bound for Aruba and jobs at the oil refinery there. Less than 200 miles away from her destination, at 11.38N, 67.42W, U 217, under the command of Kapitänleutnant Kurt Richenburg-Klinke [94], surfaced

---

93 A veteran of World War II, Bauer died in 1988 at the age of 74, after serving many years as a naval training officer.

94 On 5th June, 1943, U217 was spotted by Avenger aircraft in the North Atlantic far away

a few yards from the vessel. Crew members of U217 clambered on deck and fired five shells from the U-boat's armament at the *Seagull D*. All the shells found their target, ripping the schooner apart. These hit the top of the sail, the stern, and amidships, and killed the captain and two passengers. Four lifeboats were lowered, but two capsized. Fortunately, the hubbub was spotted by an Allied aircraft passing overhead. The U-boat crew also saw the plane and scrambled down the hatch into the body of the U-boat, which submerged immediately. Although depth charges were dropped by the plane, the U-boat escaped.

Immediately a radio alert went out to shipping in the area to go to the assistance of the *Seagull D*. Several vessels in the area converged on the scene. Many of the hapless passengers were rescued, but many others were lost, primarily because they could not swim. Survivors were picked up by different boats. Some survivors were taken to Curaçao, while others were taken to Venezuela depending on the destination of the particular boat. It took two days before the list of survivors could be compiled and released in St. Vincent. The final statistics were 3 killed by shells, ten rescued from the sea, 35 in lifeboats reached land safely and 19 drowned. Dr. Adams writes:

> (T)here was great sadness when the news spread through St. Vincent and the Grenadines. Not only had many nationals lost their loved ones, but in many cases, the sole breadwinner of some families was also lost, with serious economic consequences. The result of this tragedy, as may be expected, affected the lives of many families for many years after.[95]

The only other schooner loss recorded near to Grenada is that of the *Thomas B. Schall*, which disappeared in the area west of Trinidad on 14th December, 1942. [96]

The *British Consul*, the oil tanker torpedoed by Achilles in Port-of-Spain harbour, and again made serviceable, and the *Empire Cloud*, a British freighter of 6,000 tons, were both sunk south-south west of Grenada on the same day as the *Seagull D*, 18th August, 1942. The *Christian Kapmann*, a small Canadian freighter carrying sugar, was sunk east of Grenada on 2nd November, 1942, and the *Leda*, a large Panamanian

---

from its usual Caribbean haunts. U217 was attacked until it went down with Kapitänleutnant Richenburg-Klinke and all the crew. The captain was 26 years old.

95   Adams (2002), P. 172
96   Kelshall, P. 245

tanker of 8,500 tons, was also sunk south-south west of Grenada on 3rd November, 1942. These ships sunk near Grenada are only a few of the ships that met a frightful fate in the Caribbean area of operations.

The U-boats usually operated above water, and the U-boat crews would watch as their victims burnt, eventually sinking into the depths of the sea.

For merchant seamen, the Caribbean, usually regarded as a holiday paradise, was a place of horror. Thousands died instantly of injuries sustained when their cargo ships were torpedoed, while others died later from less serious injuries or from heatstroke, thirst, despair, drowning or being eaten by sharks. If they were lucky, merchant seamen were spared the experience their ship being torpedoed, but they lived in daily fear of this horrible experience.

VII: RESCUING SURVIVORS

The captains of U-boats often picked up men from the lifeboats of the ship they had just torpedoed. Kelshall writes:

> There are also cases on record of survivors being scrubbed and put ashore, or placed in positions where they would be quickly rescued. One survivor spent five happy days about the U126, before being passed to a Venezuelan vessel. [97]

The captains of the U-boats would often question survivors brought aboard their vessel as to the type of cargo their vessel carried and its destination.

Survivors from torpedoed ships were picked up at sea by passing boats or Barbadian, Grenadian, or Vincentian fishermen. Islanders maintained a coastal watch and rescued any small boats, life boats or improvised rafts with survivors that were sighted drifting near to their

---

97   Kelshall, P. 84

shores. Sometimes a few survivors would reach the shore through a combination of swimming, floating, and being carried by the currents. They would come out of the sea and stagger ashore, often barely alive, in a state of shock, suffering from exposure, nearly naked, badly sunburnt, starved, and virtual skeletons. Many had serious burns or were otherwise injured, and were often covered in engine oil. Sometimes some survivors were deranged. The sacrifices the islanders had to make to care for these are mostly forgotten. The survivors first had to be clothed and fed, and then cared for in the local hospital where they had to be treated and medicated. All of these services were given willingly as a service of one human being to another, the very scarce resources of the rescuers depleted without recompense. Islanders were known to give the "shirts off their backs" to strangers.

In his book *People on the Move*, Edgar Adams tells of 41 survivors from a ship torpedoed off Tobago who reached the Tobago Cays in July 1942 in a lifeboat and two rafts. They were guided to Mayreau where they were given what help was available. When the survivors set out the next day for Kingstown,

> They were all deeply appreciative of the treatment of the people of the Grenadines, some even shedding tears when they were told that the clothes they had got from the people of the Grenadines were to be returned as the people would not possibly afford to lose them. [98]

Grenada had her share of the shipwrecked, although the instances are less well documented than those rescued in some of the other islands. As the currents move west-southwest towards Grenada, survivors could come from the vessels that were sunk close to Grenada, or from vessels sunk hundreds of miles away, like the Norwegian freighter *MV Baghdad* that was sunk on 13th May, 1942, almost 400 miles north-east of Grenada. In the museum at Westerhall, there is a silver cup engraved with the date of 5th June, 1942. This cup was presented to a former headmaster of The Grenada Boys Secondary School, Jackie Grant, and his wife. [99] The inscription of the cup expresses grateful thanks from the 21 survivors of the *Baghdad* who had to wait for over a month

---

98 Adams. 2002 P. 169

99 Jackie Grant was also a cricketer of note and a Presbyterian missionary who had previously served in Africa.

before they were repatriated, during which time the Grants saw to their clothing, housing and comfort.

Larger lifeboats, like those of the *Baghdad*, could hold up to 25 people. They were usually clinker-built with straight keels, and equipped with engines. Lifeboats encountered in Grenadian waters by fishermen or other sailors would be directed to St. George's or Carriacou. If a lifeboat could not come in under its own power, the authorities were alerted to send a tug to bring it in.

In St. George's, a number of empty lifeboats, locally called "survivor launches", soon collected in the vicinity of the Treasury building where in those days there was a small beach. Other lifeboats were hauled up on the beach on the other side of the harbour. Julian Rapier and Arthur Bain both recall that a lifeboat from the *Baghdad* remained for many years drawn up on land at the far corner of the Carenage, near the pier. Ray Smith remembers that it was about 20 ft. long and had an engine. This lifeboat was later used to start boat races, and otherwise utilised by the harbour police. However, most of these sturdy craft were spoilt because they were improperly propped on shore. As a result, their keels became "hogged" or bent, and they were neglected and left to rot.

Anastasia La Guerre remembers sailors arriving in Sauteurs, having saved themselves by swimming for quite a distance. They arrived fully clothed in a blue and white uniform, and of course, wet. They could not speak English, but every one of them was so very glad to see people. She remembers that they were sent off to St. George's. Bertrand Pitt recalls that three men from a Norwegian ship torpedoed off Grenada came ashore at Pearls. These had to be taken immediately to St. George's for treatment. Janice Bain thinks that the first survivors the people of St. George's saw were sailors off of a Greek ship.

Two Germans, most likely submariners from a U-boat, were brought to C.A.O. Phillips [100], Angela Phillips' [101] father, a planter and agriculturalist who lived in Victoria. He sent for the ambulance to take them to the Colony Hospital in St. George's. While awaiting this transportation, they enjoyed the hospitality of the Phillips' house. Angela and her sister had been spared the propaganda that all Germans were fanatical, baby-

---

100   Charles Arthur Ormond Phillips
101   Mrs. Raymond Smith

eating Nazis. Thus, the girls made friends with the injured men in the way that children do, communicating without the aid of a common language. A few days later the family went into St. George's and Angela asked for permission to go and see the injured German seamen at the hospital. What a delight it must have been for these men to get a visit from these Grenadian children, who were compassionate enough to visit them as strangers who would have no other visitors, even though the children must have known that they were the enemy! Later on, the two seamen were shipped out of Grenada for destinations unknown.

On occasions when a lifeboat reached land, and depending on where they landed and on how well the occupants were, they were either offered hospitality and housed by Grenadians, or directed to St. George's, if necessary with a guide or in an ambulance. When transport was available, the survivors of friendly nations were sometimes shipped to one of the survivor camps in Trinidad and then repatriated. More often, they remained in Grenada for months and were supported by the community as the Government had asked the population to do. Survivors arriving near Grenville not *in extremis* would be taken to Dorabella Shears home. She would either give them temporary shelter, or ask someone else in the community to take them in.

Edward Kent recalls:

> One evening about seven o'clock, one of the Miss Forteau (who manned the telephone exchange) called to ask if I had heard of the lifeboat with the survivors that had come in. They were housed at Mr. Croppers Guest House, the only one on the island. I took some fruit and half of a roasted chicken, which was all we had, and went to them. There I met nine Norwegian sailors whose ship had been torpedoed to the northeast of Barbados....They had with them a sailor who had been standing on deck when the torpedo struck and whose feet were badly burnt. He was in the Princess Royal Hospital, but the next day the commissioner sent them off in their lifeboat (with a guide) to St. George's where the injured sailor could get better medical attention.[102]

As a child living in Carriacou, Robby Rowley recalls that an English vessel was torpedoed off Carriacou. His father, Redvers Rowley, was the District Officer for Carriacou at the time. He brought two survivors

---

102 Kent, P. 74-75

from this vessel to stay in the Rowley home until they could be sent back to Grenada, and then to England. The launch in which the survivors had arrived was given to the Government of Grenada, and remained in service for many years.

In Grenada, all shipwreck victims known to be German or those suspected of being enemies, were taken to the Colony Hospital for any treatment necessary and then held for further "observation" until they could be handed over to British authorities and transported to Trinidad. Similar protocol was observed elsewhere.

In addition to the survivors, dead bodies washed up on Pearls and Telescope beaches. Harry Ogilvie remembers that bodies of white people washed up near Telescope beach in St. Andrew's. Bertrand Pitt recalls a tale told of some sailors trying to reach shore outside of Pearls, but they were taken off the coast by the current and eventually drowned before they could be rescued. According to Nellie Donovan, bodies also washed up in Black Bay, but she did not know what happened to them.

The body of an unknown British soldier rests in the grandeur of Westminster Abbey in London. In Carriacou there is a grave of an unknown British Sailor, whose body was washed ashore. The simple grave is on Hillsborough's hillside cemetery amongst the graves of the valiant local people of Carriacou. The view from the site is spectacular, with Grenada and the small islets of the Grenada Grenadines set like stepping stones across the sparkling Caribbean Sea to Grenada. The dark shapes of the U-boats lurking beneath the sea that claimed this sailor's life, and the lives of so many others, are gone now, leaving only the beauty and tranquillity of the magnificence of the seascape. The grave of the unknown sailor is still a focus for Remembrance Day celebrations in the sister isle and is maintained by Britain.

VIII: WHAT THE SEA BROUGHT IN

Local seamen reported that they encountered all sorts of debris as they plied their way through Caribbean waters, including general wreckage that occasionally contained a recognisable piece of a U-boat, ships' masts, tables, chairs, mattresses, entire lifeboats, and anything from sunken ships that could float. Among the items were certain foodstuffs. On his passages to and from Trinidad, Captain Mark Hall would see and haul aboard his boat drums of edible oil, which he would distribute in Grenville to family and friends. The tides would also wash this debris on to the shores of the islands.

Grenadians called this bounty from the sea "come-ashore". Items remembered as "come-ashore" in Grenada included chunks of raw rubber from Brazil, about 1½ feet in diameter. These were of very little use except as curiosities because the rubber was sticky and would flake if there was any attempt to slice the cube, since this rubber was only partially processed. However, usable rubber in bales also washed ashore and was put to many uses, including the replacement of shoe soles. Other "come-ashore" items included lumber, flour, jam in tins, butter in tins, cans of *Cow and Gate* and other tinned milk, corned beef and other canned foods, and peanut oil and cod-liver oil. These oils were used for cooking, even if their smell and taste were unfamiliar and unpleasant, especially if the oil became rancid. Bags of flour were salvageable because the outer layer that got wet formed an impervious case around the balance. All that was necessary was to break through the casing to get at the good part. Occasionally shoes and clothing would also appear. With the useful "come-ashore" came detritus and debris, which was picked over for anything that might be useful, including wood for cooking fires.

"Come-ashore" washed up on the beaches in Sauteurs, Pearls, Telescope, Grenville, Victoria and Windward in Carriacou and occasionally in Gouyave, Black Bay and Moliniere and other beaches. All "come ashore" was supposed to be reported to the Government, but this instruction was "honoured in the breach". People living on or near the coast made it a habit to get up early to see what was on the beach or take their boats out to pick up goods floating on the sea. Judith Parke remembers a fisherman called Mr. De Peiza, who lived in Marian but fished from Calivigny. De Peiza used to find and collect items washed ashore in that bay. Among the items he found were tins of corned beef, cooking oil, and lumber. Bertrand Pitt identifies the most popular areas on the St. Andrew's coastline where people would most likely find useful items as Grenville "bay side" and Pearls, especially near the end of the airport runway.

Once Cecil Edwards's father and some of his friends found seven tins on Shenda and Moliniere rocks-and divided them among themselves. Some were brought up to the village where Cecil lived with his family. After much protestation by the women, who feared the tins might be bombs, they were opened. Luckily for all in the vicinity, the punctures produced streams of olive oil, called "sweet oil". The oil was shared out and used for cooking.

The foodstuff and material obtained from "come-ashore" helped to ease the shortages and economic hardship of the very poor. Searching for "come-ashore" also added a little bit of excitement to life, as one never knew what a morning's search of the shore would bring. However, all "come-ashore" was not innocuous.

IX: A DANGEROUS HARVEST

Apart from the U-boats, shipping and Caribbean coastal populations had to face the danger of mines torn loose from marine minefields. Mines live shells and other explosives washed up on the shores and could be mistaken for harmless "come ashore".

On one occasion, some drums washed ashore near the Church of England in Sauteurs. Carlyle John recalls that an attempt to retrieve them was made by Vernon Springler, a fisherman better known as "Tuppence". A drum exploded, injuring him to the extent that he had to be taken for medical attention. In addition, the entire town of Sauteurs was polluted with a very bad smell. "Tuppence" had come across floating mines of some sort, thankfully largely non-functional.

The U.S. Navy, needing to secure the Gulf of Paria in Trinidad, determined that the only way to do this was to string mines across the Dragon's Mouth and the Serpent's Mouth – the two main entrances to the Gulf. The British advised that the current was too strong, and that the mines would not only break loose but would be ineffective because the strong current would keep the mines lying at an angle to their moorings, too far underwater to be useful. In spite of proving this point by passing several vessels drawing as much as 22 feet over some mines, the Americans went ahead. As predicted, within three months the mines began breaking off and became a danger to shipping. Kelshall observes:

> In theory the mines should have deactivated themselves as soon as their cables parted, but this was not taking place and the Dragon's Mouth area was being rocked by frequent shattering explosions as some of the mines ran into the rocky shoreline. However, only a small proportion of the drifting mines were hitting the shoreline, the majority were being carried out of the Bocas by the current and up into the Caribbean.[103]

Of the 35 mines that got loose from their moorings, only some of these were spotted by the aircraft diverted to search for and destroy them. Others caused tragedies. Victims of these floating mines included the British Freighter *SS Pan Gulf*, struck off Cedros Point in Trinidad, and the British freighter *SS Athelbrae*, which hit a mine in the Serpent's Mouth. It is also believed that it was a floating mine that blew the bow off the American freighter *Frederick R. Kellog*. Kelshall believes that a "high proportion" of the Caribbean schooners that went missing were not sunk by U-boats, but were struck by these floating mines.[104]

The Americans also laid mines in the approaches to San Cristobal Harbour, Panama, and in other locations, but these would not have drifted into the Caribbean. They were nevertheless a danger to shipping in the area and at least one ship was damaged by the mines that broke loose from this harbour.

The USA was not the only country to lay mines. French *Briquette* mines cut loose from the harbour at Dakar, Senegal and floated across the Atlantic into the Caribbean. The disaster in Carriacou shortly after the war is known to have been caused by one of these mines.[105] German submarines also laid mines. Markworth in U66 had laid mines in Castries Harbour in July 1942, many of them remaining active until they were discovered and removed by a minesweeping operation in 1944. In the middle of October 1943, Becker in U218 laid twelve German underwater mines in the area outside of the Third Boca in Trinidad. He also laid a minefield in the harbour of San Juan, Puerto Rico in 1944, but this was discovered by the U.S. Navy's large minesweeping operation which covered all the Caribbean harbours before any damage was done.

---

103   Kelshall, P. 244
104   Kelshall, P. 122
105   See Section VI Part III

X: THE U-BOATS DISPLAY THEMSELVES

Particularly after the attack on the *Lady Nelson* in Castries, the population of Grenada became very aware of the U-boats. On the other hand, the U-boat captains took full advantage of the powerlessness of the smaller islands to flaunt their command of the sea. Bridget Brereton says with reference to Trinidad and the Eastern Caribbean that the waters around Trinidad:

> Were infested with German submarines and they sometimes surfaced and shelled reconnaissance planes based on the island; occasionally they landed at lonely beaches in the eastern Caribbean to exercise their crews.[106]

Kelshall states that

> There were places where the U-boats could call and buy fresh provisions from natives who didn't understand and didn't care what the war was all about. Some of these visits are reputed to have involved the provision of female company as well.[107]

The U-boats went everywhere unmolested during 1942 and 1943, and possible encounters with the U-boats added to the normal hazards of travel by sea. The U-boat captains ranged at will and made sure that Caribbean populations knew that there was nothing to stop them going where they wanted to go. It was as if the U-boats wanted people to see them and know they were there! The submarine captains also took their vessels close to the islands, surfacing in the broad daylight near the chief towns which were usually situated on harbours, not so much looking for prey as for reconnaissance, or just because they could.

In Grenada, U-boats were frequently seen at different points around the island by the observant. Nellie Donovan recalls seeing the periscope of a submarine off Ross's Point when she went to visit her fiancé, Herbert Payne, who was the Commanding Officer of the soldiers stationed at Ross' Point. The submarines were frequently sighted and heard by

---

106 Brereton, P. 191

107 Kelshall, P. 85

the soldiers at this station. Cecil Harris remembered how the soldiers used to "tremble" at the sight of the U-boats, and how the U-boats seemed to be "teasing them" as they sailed along the coast but they never did anything but make sure they were seen. The young soldiers could only watch in readiness, as the order to fire was never given. The soldiers never found out that their battery had very little ammunition, and in any case, each U-boat was powerfully armed, and the guns of the shore batteries would certainly be no match for the retribution from the powerful guns of the U-boat, which could do immense damage on shore.

Some of the windows of the Phillips' house on Upper Grenville Street overlooked the sea. From these windows Margaret Phillip used to observe U-boats just offshore. One Saturday during the war, Gunny Swapp was sent from his home in Dry River to Grenville "to make messages". Although only a small boy at the time, he remembers seeing an agitated crowd looking into the waters of the harbour. When he looked too, he saw a U-boat surfaced in the harbour. He recalls distinctly that it stayed there for quite some time before it left the harbour, submerging as it went. He also remembers that he was not frightened because of "all the people around him".

On another occasion, several people saw a U-boat surface in the Carenage, remain there for quite a long time ostensibly so enough people would see it, and then leave. Cecil Edwards rushed to the harbour to see the U-boat, but he was in time only to see the departing periscope. Irvine Redhead, the harbour master was accustomed to take the Government launch *Doris* around St. George's Harbour to see that all was well. After this incident, he felt that he had to make a response to this sighting, and made an unscheduled tour of the Harbour armed with a shotgun in search of the U-boat. Perhaps fortunately, this exercise was futile.

Louie Steele lived at Woodford Estate with her father in a house overlooking Halifax Harbour. Her niece, Margaret Peebles, has documented that:

> During the Second World War, Louie remembered a German submarine arrived in the bay and surfaced to allow the crew to breathe some fresh air and swim in the tropical waters. Fortunately the crewmen stayed on the beach down below and did not venture further ashore, while those at the

> house watched quietly from the parapet until the submarine departed several hours later.[108]

There were two brothers who lived near the Carenage who liked to go out to fish at night in their small boat in St. George's Harbour. At this time, no fisherman went out too far, because the outboard motor had not yet been introduced into Grenada. In any case, these two were fishing for fun, and there was usually plenty of fish to be had near the shore. On one occasion, these brothers spotted a U-Boat in the harbour. There is no doubt that the U-boat also spotted them!

The U-boats were not only seen in the day, but they also operated near to the shore at night. Michael Humfrey writes:

> There were rumours among the fishermen that submarine crews sometime came ashore after dark to stretch their legs and gather coconuts; and one moonlit night, a neighbour of ours who had a beach cottage at the northern point of Grenada actually saw the ugly silhouette of a U-boat pass close inshore, its crew relaxing on deck.[109]

Many are the tales of German sailors from U-boats attending the cinema in St. George's, or seeking the company and solace of Grenadian women in the villages outside of St. George's.. There are rumours that German sailors were given "shore leave" from their submarines and came ashore in small boats near Ft. Jeudy, Woburn and Calivigny, in search of "rum and girls". A woman on the Carenage is supposed to have had several children for German sailors, identified by their light skin.

People living in St. David's often heard the distinctive high pitched whine of the generators of the U-boats as they recharged their batteries on the surface just off the coastline or near the off-shore islands. Those living at high elevations in St. David's could also observe U-boats on the surface travelling along Grenada's coastline. Fishermen commonly told of hearing strange noises and unexplained sounds while they fished at night. Austin Hughes is one of many persons who heard the motors of submarines coming into St. George's Harbour, saw their lights off-shore and heard them as they recharged their batteries on the surface.

The U-boats made a haven out of the Grenadine islands, including the Tobago Cays. This archipelago of over 120 islands, cays, rocks

---

108   Peebles, P. 29
109   Humfrey, P. 84

and atolls stretches over the 70 miles that separate the main islands of Grenada and St. Vincent. The sailing in the area is magnificent, and bathing is superb. Best of all for the U-boats, the islands and cays were either uninhabited or very sparsely populated. Only Carriacou, Petite Martinique, Union, Canouan, Mayreau, and Bequia had populations during the war. The Grenadines were the perfect place for the U-boats to surface to recharge their batteries, refresh air in the U-boat, and give their crew the opportunity to exercise and swim in the fresh air and sunlight and pristine tropical waters, to fish and help themselves to coconuts. Angela Phillips used to spend summer holidays at Green and Sandy Islands off the north coast of Grenada. At nights during the war her family could hear the generators of the U-boats as they re-charged their engines nearby.

From this distance in time, these tales seem to belong in the realm of legends, except that there was a time in the war when the Germans thought themselves invincible, and captains saw little danger in allowing their crew to disembark to enjoy themselves on an island under the British flag. This was especially easy as not all Grenadians understood that they should not consort with Germans, especially if there was something in it for them.

The captains of schooners and sloops travelling between Grenada and Trinidad would often encounter U-boats during the journey. Cosmos Cape recalls travelling to Trinidad in 1941 on business for his father in a sloop with a Johnson Seahorse outboard motor. During the passage, which usually lasted 1 ½ days, those on board noticed a U-boat "quite close", but sailed on unmolested. It was infrequent that schooners were stopped and challenged, and it was even rarer that the U-boat sunk an inter-island schooner, for the objective of the U-boats was to sink large ships. Success for the U-boats was measured in tonnage of ships sunk, and the inter-island vessels were insignificant in this regard. Later in the war, however, schooners were challenged and sunk if the U-boat captains suspected that the schooners were ferrying workers and goods between the smaller islands and the larger ones and thus defeating the blockade aimed at completely cutting commerce between the allies and their Caribbean colonies.

There is no record of any Grenadian boat being sunk by a U-boat, although U-boats would sometimes stop fishing sloops and commandeer their catch of fresh fish, letting the sloop and the crew depart otherwise unharmed. Cecil Edwards remembers hearing a tale of a German U-boat stopping a schooner bound for Guyana. The U-boat captain asked for news of a particular schooner, which was in fact the same boat he had stopped. The quick-thinking schooner captain told the U-boat captain that the boat he wanted was following his own. He got safely to Guyana, and lost no time in repainting his boat and changing the name.

Then there is an oft repeated tale of an incident of a U-boat captain stopping a small steamer on the way to Trinidad. The steamer captain was warned that his boat was really too small to waste a torpedo on, but if he continued to ply between the islands, the next time he was caught he would be sunk.[110] The U-boat captain then gave the captain of the schooner two tickets to the Roxy Cinema in Port-of-Spain, telling him that he should go and see that show, because it was very good.[111] Alimenta Bishop thinks this is the same steamer that brought her and her baby girl Ann to Grenada for the birth of her second child, as this same tale was related to her by the captain. Mrs. Bishop and her baby daughter had travelled from Aruba to Trinidad on a large vessel, transferring for the passage to Grenada to this smaller steamer which had been sent to Trinidad by the Competent Authority in Grenada for flour.

XI: CONSEQUENCES OF U-BOAT ACTIVITY IN THE CARIBBEAN

What were the consequences of U-boat activity in the Caribbean? The consequences were exactly those that the German High Command had envisioned.

Firstly, in England, the loss of shipments of oil from the Caribbean caused

---

110 Kelshall, P. 83

111 This tale has even found its way onto Caribbean Fiction. In *Prospero's Daughter* by Elizabeth Nunez, it says on page 170: "Then one night under cover of starless sky, a German U-boat sneaked into Port-of-Spain harbour. Days later, a German officer produced a movie stub from the Globe, a cinema in the heart of the city. He had eaten well, gone to the movies, slept with a woman, he boasted. "All done under the eyes of English and American soldiers."

Britain's oil reserves to drop to danger level. The war could not go on if the RAF and Britain's navy could not get oil and fuel for the vessels and aircraft transporting war materiel to British and Allied troops fighting in Europe. The RAF also needed an unlimited supply of aviation fuel to maintain attacks on Germany and to protect Britain. Although no British campaign anywhere in the world had to be cancelled, the concern was great, and some operations were slowed down, others curtailed or some postponed. Additionally, the loss of tankers meant that oil from the refineries could not be taken away as speedily as it was produced, and production fell to only a small percentage of capacity.

Secondly, air power was the single most decisive military factor in the Second World War, and the seriousness of the inability to build aircraft fast enough due to a curtailment of shipments of bauxite was of crucial strategic significance.

Thirdly, extremely effective economic warfare was wreaked in the Caribbean area, with ships carrying oil, bauxite and food supplies from the Caribbean to Europe and America falling victims to the German U-boats. The supply line of raw materials to Britain was almost cut, and the colonies were temporarily deprived of their markets. Moreover, very little food and other manufactured items could be exported to the Caribbean.

The curtailment of trade was devastating to Caribbean economies as well as to the United Kingdom and the USA. After the initial losses of cargoes of sugar and other food items from the Caribbean to England, these shipments were strictly curtailed, leading to rationing of sugar and imported items such as coffee in England. To assist its colonies, Britain purchased the entire British Caribbean sugar crop at fixed prices even though the sugar could not readily be shipped. But Britain did not buy out other products such as cocoa and nutmegs which were the Grenadian farmers' livelihood. The hardship caused to the planter, merchant and small farmer alike in Grenada when shipments of nutmeg and cocoa from Grenada ceased has previously been mentioned.

Freighters travelling from Britain carrying food, manufactured goods and other necessities for the Caribbean were likewise torpedoed and the residents of the Caribbean were very soon became discomfited as ships bringing goods to the islands came few and far between and only to the larger ports. From the beginning of the war there were certain imported items the Caribbean people would not see again for the balance of the war and for a few years after that as well. Some islands that depended heavily on food imports faced a food crisis.

The United States, too, felt a squeeze in their food supply when they were not able to import sugar, fruits, and beef from South America for domestic consumption. They also missed the leather that they imported from that destination. The use of oil for domestic consumption was curtailed. But the first area to suffer was the Caribbean.

Perhaps the most important consequence for the Allies was that at the height of the activity, the command of the seas was temporarily lost, with the transatlantic shipping lanes closed to all but the most vital shipping. Communications between Europe and the Caribbean were severely disrupted, and the sea lanes through the Caribbean almost unusable. The U-boat activity also presented a great and unexpected threat to the United States, whose shores were for a period entirely vulnerable to this type of warfare.

Finally, and most importantly, thousands of lives of merchant sea men and passengers were lost. Admiral Herbert Hoover is quoted as saying that the U-boats "had a field day down there. About all we are doing was picking up survivors."[112] Damage is often counted in terms of tonnage of shipping sunk, but there was a tremendous loss of life among merchant sea men in Caribbean waters. More than 7,000 sea men died, among them were 74 from Barbados. The names of these Barbadians are inscribed on a memorial in Bridgetown's Military Cemetery.

On 26th February 1942, the Trinidad Leaseholds oil tanker, *La Carriere*, was hit and sunk south of the Mona Passage, with the loss of 16 Trinidadian lives. There was loss of life amongst civilians as well. When the *Lady Nelson* was torpedoed in Castries harbour, 15 deck passengers were killed, fourteen of whom were Barbadian. Three crew members also perished. When the *Lady Hawkins* went down in the North Atlantic, on 20th January, 1942, several Trinidadians and one Grenadian were aboard. The Grenadian was Robert Charles, Bertrand Pitt's uncle-in-law, who had been living in New York with his wife Gladys (née Pitt) from Beaulieu. This couple had decided to return to Grenada, and Gladys was sent ahead, arriving a few months before Robert's passage. When she was informed that her husband had been lost, she was absolutely devastated at the news, and died shortly after.

---

112 Hoover, quoted in Baptiste 1988, P. 143

It must be realised that Grenada and the smaller Caribbean islands could have been captured by Germany very easily, if this was on the agenda of Germany. In the larger islands it was possible that much mischief could have been done by Germans. Kelshall points out that:

> With the extraordinary freedom the U-boats enjoyed, they could have easily landed teams of trained saboteurs and commandos and taken them off again. Had these tactics been employed in conjunction with the massive onslaught on shipping between May and September 1942, untold chaos would have been caused in inadequately defended islands.[113]

XII: SCHOONERS TO THE RESCUE

Just before the war, small to medium sized freighters popularly referred to as "iron boats" had made their appearance in the Eastern and Southern Caribbean. They plied regularly between the islands, Europe and America, bringing manufactured goods and processed food directly to the island ports, then returning to their home ports with the raw materials and produce of the Caribbean. But after 1942, U-boat activity in the Caribbean Sea put an end to direct shipping to the smaller islands by these small freighters. In the Eastern and Southern Caribbean, larger ships would now come to Trinidad and Barbados only and this only when escorted by warships. The smaller islands had to find a way to obtain the supplies they needed for survival. This they did by applying an age-old solution to this new problem.

Woodville Marshall [114] defines entrepôt as a trans-shipment point for supplying goods and services derived from a metropole to a sub-region. Such entrepôt trading with schooners providing the feeder function to the islands was historic in the Caribbean, which only slowed with the advent of the small "iron boats". Once more, locally-built wooden boats stepped in to provide an essential part of a revived entrepôt trade made necessary by the exigencies of war. The word "schooner" is often used as a generic term to refer to:

---

113 Kelshall, P. 265
114 Marshall, P. 4-5

> The array of small trading ships – schooners, sloops, motor vessels – that have carried (and still carry) cargoes, passengers, packages, messages and news up and down the island chain." [115]

Most islanders distinguish between these boats. Schooners are large wooden sailing vessels, characterised by fore and aft sails on two or more masts with the forward mast being shorter or the same height as the rear mast. Although weighing as much as 100 tons, schooners are sleek and fast.

> They were fine, well-built ships, strong and seaworthy to survive the unpredictable storms which frequently blew up in the south Caribbean, and in a matter of two or three minutes could transform a calm sea into a maelstrom of deep troughs and flying water. Their graceful balanced lines showed the unmistakable traces of their thoroughbred ancestry in the 19th century boatyards of Marblehead and Maine. [116]

Schooners have been built in the Eastern Caribbean for centuries. In Grenada, Scottish shipwrights were brought to and settled in Windward, Carriacou to build and maintain a fleet of such vessels. The St. Vincent Grenadines have also produced hundreds of these beautiful craft over the years. A sloop is a smaller one-masted vessel with a fore-and-aft rig. Quite capable of travelling wherever the schooners went, their small size made them more suitable for transportation of people or trading between nearby islands than for the entrepôt trade. Finally, a motor vessel is entirely dependent on an engine, having no sail. Its shape is completely different from schooners and sloops. Motor vessels are essentially launches used for coastal travel.

As the demand for a flotilla of the schooners grew, ship-building proceeded in earnest. At the end of the war, Grenada had 27 registered schooners and 46 registered sloops. Of these, 32 had been built during the war, or in 1939.[117] Schooners and sloops were essentially sailing

---

115   Marshall, P. 4

116   Humfrey, P. 88

117   After the war, the wooden boats so expertly built in Grenada and in St. Vincent and the Grenadines were gradually replaced by steamers and small steel-clad boats. Edgar Adams recalls that some of these were second-hand boats bought by sea captains of the two islands from enterprises whose boats had been holed up in Swedish and other European ports during the war, and were now out of date, but eminently suitable for the inter-island trade in the Windward Islands. In spite of this and other drawbacks, such as the difficulties of getting insurance for wooden boats, local schooners are still in evidence in the harbours of Grenada

boats, but because of the need for increased speed and reliability, engines were added to the boats. Thus almost all the old schooners were converted into auxiliary schooners, capable of sailing under canvas, by using their engines, or using both. The new boats were all built to travel under both sail and engine. Most local boats carried no navigational or radio equipment, and once they were out of sight, there was no contact with the vessel until it returned, unless it was sighted by another boat or aircraft. Because engines did break down, and boats could be blown off course or becalmed, schooners usually travelled with at least one day's supply of emergency rations. Because of the uncertainties of seafaring, when local vessels put out to sea, there was always uncertainty when they would return.

Notwithstanding the dangers of U-boat activity and the natural hazards of the sea, islanders were anxious to go to work in Trinidad and the "ABC" islands – the three Dutch islands of Aruba, Curaçao and Bonaire. Aruba and Curaçao both had large oil refineries and as such, the migration was mainly to these, and not to Bonaire. Schooners and sloops transported passengers seeking work. They carried people who needed to travel to see their families, for business, to meet a liner to go out of the Caribbean, and even to go to school. They carried local foodstuffs for sale in Trinidad and Barbados, and brought back from those islands much-needed scarce commodities.

At first, the schooners and sloops travelled between the islands at will. However, when the importance to the smaller islands of the entrepôt trading was recognised by Germany, the U-boats began to take more frequent action against the schooners, resulting in bereavements, disruptions, privations, financial losses and shortages, which made the islanders restive, anxious and uncomfortable.

The threat became so serious that the schooners began to move mostly at high noon, when the U-boats generally hid underwater, and would seldom sail at night when the U-boats were on the prowl above water. However, the U-boats grew so confident that they would attack in the day as well as at night. The *Mona Marie* and the *Seagull D* were two schooners that were shelled in daylight.

---

and Carriacou, and new ones are occasionally built in Windward and in Carriacou. Some captains prefer their beautiful handcrafted vessels because they are so well suited both to the trade and to the marine conditions that they cope with on every voyage. See Adams 1996 Pg 136-137.

No one is absolutely sure how many schooners were destroyed by U-boat action. To the Germans, schooners were of little account, except as part of the entrepôt trade. Each schooner probably weighed less than 100 tons, while freighters and tankers could weigh hundreds of tons. As Germany and the Allies counted shipping losses by the tons, the schooners were too insignificant in weight to record. The U-boat captain might make a notation in his deck log of a schooner sunk, but he might not do even this. The indications are that a large number of schooners were sunk, and possibly only a half recorded.[118] Known to be sunk on the 3rd June, 1942 by a U-boat was the 80-ton *Lillian*. Another schooner was sunk by Becker in U218 on 12th November, 1943, but Becker's records do not cite the boat's name. Warren Alleyne records the names of five other Barbadian schooners sunk by the U-boats: *Vivian P. Smith, Florence M. Douglas, Glarier, Harvard and Gilbert B. Walters*. No schooner from Grenada is recorded as having been lost through enemy action, and the same may be said of St. Vincent, with the exception of the *Seagull D*. However, in *The Times* of St. Vincent of 19th August, 1944, there is the statement that:

> Quite a few vessels have been reported missing for the past few months; one or two have reached the Venezuelan coast while nothing has been heard of the others...

Whether any Windward Islands schooners were sunk by the U-boats or not, or were lost through misadventure, the additional risk the sailors had to face with the U-boat presence in the Caribbean was a risk they could have done without, faced as they were with the ever-present ordinary seafaring risks. Some sailors experienced so much stress with the advent of the U-boats that they were left with severe health problems for the rest of their lives.

The schooners, treated so casually by those recording vessel losses, meant everything to their owners. A schooner was usually the owner's lifetime investment and, in addition, it carried a valuable cargo of food or goods, on which the islanders depended. If the schooner trade ceased, there would be no other way of getting the essentials of life to the islanders. Schooners and their role in keeping the islanders supplied with essentials during the war only became an important concern to the Allies when, with the increasing losses of schooners, there was the

---

118 Kelshall, P. 122 and Adams, 2002 P. 171

threat that the schooner trade might be disrupted because the captains of the schooners became increasingly unwilling to risk their lives and boats in further trading operations.

Fearing the complete cessation of the entrepôt trade, the West Indies Schooner Pool Authority was formed. This was said to be a suggestion from President Roosevelt himself. Under the auspices of the Anglo-American Caribbean Agreement, the schooner pool was officially designated in August 1942, and began operations in August 1943. This was long after the worst of the U-Boat activity, but it was "better late than never."

The Pool operated from Puerto Rico, Barbados and Trinidad. The trade, once informal and individualistic, would now be organised. Cargo vessels would offload essential imports in ports in these islands, which would then be stockpiled. Privately operated schooners would be organised to transport these goods to various ports in the Leeward and Windward Islands. Vessels in the pool were advised where and when to travel to bring back food and supplies, and the vessels sailed in convoy with navy protection. To identify local shipping to the Allied patrols, craft carried identification letters and numerals that had to be displayed prominently on the sails, bow and stern of the vessel. Boats from the Windward Islands carried the letter "W" and those from the Leeward Islands the letter "L". Thus the *Harriet Whittaker*, the boat belonging to the U.S. Consulate in Grenada and named after the wife of the Consul, had the identification code W257, and the *Emmanuel Florence* was W256. After the attack on the *Cornwallis*, the schooners in convoy were provided with the additional protection of air cover.

Although it had its problems, the schooner pool greatly alleviated the supply situation in the Eastern Caribbean. There were some inefficiencies causing uneven distribution and short supplies from time to time in some islands but, undeniably, the Schooner Pool ensured that the islands survived through the most dangerous months of the war. Generally, it was more efficient because the most urgent supplies had precedence, and the distribution of cargo received in the larger islands for trans-shipment to the smaller ones was accelerated. It is estimated that between 75-80 percent of all small vessels that operated regularly in inter-island trade joined the scheme. Fullberg-Sholberg estimates that in July 1943 this amounted to 84 ships with a total load capacity of 7,000 tons.[119]

Since most of the schooners came from Barbados, the headquarters of the Schooner Pool was set up in that island. Operations were under the command of Captain H. S. Threw, an American headquartered in Barbados. Although the Schooner Pool was in operation, several Grenadian schooners and sloops as well as boats from other islands, continued their lone runs to Trinidad and other destinations. Sea men still had to make their livelihood by operating boats and crewing for them, and the work for the schooner pool was simply not enough to make ends meet.

There is no record of a ship in the Schooner Pool being attacked, but the crews and their families of schooners were comforted by the death benefits provided with membership in the Pool. A fee was charged for each ton of cargo carried. The money went into a fund which would be used to pay a bounty to the families of the crew members who lost their lives if a vessel in the Pool was lost by enemy action. The Schooner Pool outlasted the war in the form of the Schooner Owners Association.

---

119 Fullberg-Sholberg, P. 111

PART IV:

A SCRAMBLE TO DEFEND THE CARIBBEAN

I: VESSELS FOR DEFENCE

While Grenadians and other citizens of the Caribbean islands tried to adjust to the changes that the war made to their lives and to the dangerous Caribbean Sea, the United Kingdom and the United States scrambled to find counter-measures to the U-boat deployment in the Caribbean and the havoc they were causing. One of the first measures was to change the shipping routes, but the new routes were quickly discovered. The blackouts were another measure, as was the zigzag route that vessels employed to make attack by submarines that more difficult. But it was not until 18 months after the first wave of U-boats arrived that the tide began to turn. The Allies were able to deploy adequate aircraft for anti-submarine manoeuvres and sufficient escort vessels, equipped with devices to detect U-boats so that the convoys were sufficiently protected. In a short time the Allies regained command of the Caribbean Sea.

As a part of the defence of the Caribbean, the Allies installed anti-submarine measures from the Canadian coast to the southern boundary of Brazil and in the Caribbean, mainly around Trinidad. As long as there was a threat of U-boat activity in the Caribbean theatre, these anti-submarine forces stayed in place to protect the coastlines of North, Central and most of South America and the Caribbean. Therefore, Germany kept submarines in the Caribbean long after the U-boats were truly effective to ensure that these anti-submarine forces stayed tied up in the Caribbean theatre and were unavailable for use in the Atlantic or other crucial zones.

As soon as was possible, additional battleships and escort vessels were sent to the Allied command in the Caribbean to augment those already deployed for convoy duty. Increased protection for the convoys meant that the much needed oil and raw materials could once more flow freely to Britain. The suppression of submarine activity was the task of the Royal Navy motor torpedo boats or "MTVs" manned partly by British Navy personnel and partly by the Trinidad Royal Navy Volunteer Reserve (TRNVR).

The MTVs were later replaced by the armed Hudson motor launches.

II: AIRCRAFT

A variety of aircraft for submarine spotting and destruction also arrived in the Caribbean, and were mainly stationed in Trinidad. The OS2N "Kingfisher" aircraft, a twin-seat, single-engine floatplane, was the first made available for the Caribbean area. Although this aircraft could not carry a bomb payload large enough to sink a U-boat, the "Kingfishers" proved useful because, just by being there, they often caused a submarine to dive, and in many cases a merchant ship was saved in this way.

Two squadrons of US Navy Catalina PBY flying boats arrived in Trinidad shortly after one another and soon after the "Kingfishers". They had a very long range, and could take a lot of punishment. They were equipped with MAD gear[120] for detecting submerged objects, and sonobuoys [121], which were to prove to be the ultimate anti-submarine detection device. The Catalinas would prove a serious threat to the U-boats.

The Army Air Corps used "The Flying Fortresses" — Douglas B18 Digby Bombers. Although ungainly and slow, this plane could carry a weapon load of 6,000 pounds and had radar in the nose. It was the B18 aircraft that were to bear the brunt of the anti-submarine effort. The installation at Vieux Fort in St. Lucia was almost exclusively for B18 aircraft.

The work of the B18s was aided by a squadron of Havoc light bombers (A20As). The RAF also posted the B53 squadron of Lockheed Hudsons to the airfield at Waller Field. These aircraft were eminently suited to the job of search and destroy mission against the U-boats. The squadron was also used to train US airmen in the art of attacking U-boats. Fulfilling both missions, this squadron played a significant part in halting the U-boat activity in the Caribbean. Also operating out of Trinidad were Mariner Flying Boats, the carrier-based Grumann TBF/TBM Avenger torpedo bombers, and the B24 Liberator Bombers. The planes criss-crossed the Eastern Caribbean daily, and became a frequent sight in the Grenada's skies or off her coast. They were painted deep grey, but Grenadians remember them as being black, noisy and frightening. Other aircraft stationed in Trinidad included the P40 Curtis Tomahawk Fighters, Percival Proctors, Blackburn Sharks and Roses and Walrus Amphibians.

Without an airport until the war, Grenadians were unaccustomed to the sound of aircraft. Now aircraft flew over Grenada regularly during the day and during the night, because Grenada was on the flight path to and from the airfields in Trinidad. The aircraft often flew solo or in twos, but at times several would fly in formation quite low, making a great deal of

---

120 A *magnetic anomaly detector* (MAD) is used to detect minute variations in the earth's magnetic field. In this context, the term refers specifically to magnetometers used by military forces to detect submarines, because the mass of the submarine would create a detectable disturbance in the magnetic field.

121 A *sonobuoy* (a combination of the words sonar and buoy) is a relatively small (typically 5 inches in diameter and 3 ft long) expendable sonar system that is ejected from aircraft or ship conducting anti-submarine warfare. The buoys are ejected from aircraft in canisters and open upon impact with the water. An inflatable surface float with a radio transmitter remains on the surface for communication with the aircraft, while one or more hydrophone sensors and stabilising equipment descend below the surface to a selected depth. The buoy relays acoustic information from its hydrophone(s) via UHF/VHF radio to operators onboard the aircraft or ship.

noise. Carlyle John and Arthur Bain remember seeing a formation of at least six bombers flying over Grenada in the broad daylight. Grenadians, at first, were terrified of the planes and the noise they made, but this fear of the war planes disappeared when the population found out that the planes overflying Grenada were either American or English, and that German aircraft did not have the range to fly over Grenada.

III: ZEPPELINS

On 10th February, 1943, the US Navy Squadron ZP-51 became the first lighter-than-air unit to operate outside the continental United States. The zeppelins, which the Americans called "blimps",[122] were first moored at Edinburgh Field in Trinidad until a special new facility was opened for the zeppelins at Camden Field, south of Edinburgh Field.

The K-type airships of the squadron could carry depth charges and had devices installed that enabled the detection of submerged objects. The zeppelins flew low and slow, and were the ideal craft to spot a U-boat trying to dodge detection. One hundred and fifty "blimps" were built by the Goodyear Company of the USA for use as submarine spotters. It was claimed that no US vessel was attacked by a U-boat while in convoy escorted by these airships.

Zeppelins became a frequent sight in Grenada, flying sometimes so low that the craft seemed to brush the treetops and those on the ground could see and wave at the crew. The zeppelins frequented Grenada's east coast, searching for U-boats hiding near the small offshore islands. Glovers Island, Calivigny Island and Hog Island were favourite haunts of U-boats needing to re-charge their batteries and to refresh their air. They would hide near these islands in wait of a hapless ship on the way to meet a convoy in Trinidad or Barbados, or a vessel chancing an unescorted journey. The zeppelins would also search the many deep narrow deserted bays, characteristic of Grenada's east coast, for U-boats on the surface charging their batteries.

---

122 Developed in Germany by Count Ferdinand von Zeppelin, these lighter-than-air airships had been used for air transport in the 1920s and 1930s.

On Grenada's west coast, the zeppelins would search the deserted bays on the way to the Grenadines where the U-boats tended to surface to change their air, charge their batteries and to give their crew some sea, sand and sun. Zeppelins were frequently seen in Carriacou as they passed up the chain of the Grenadines searching for U-boats. Godwin Brathwaite remembers seeing one east of Mount Pleasant in Carriacou. Bonace, his wife, remembers being amazed at the sight of a zeppelin circling Carriacou at a very low altitude. The zeppelins would also travel north to search the waters near Barbados and the other areas of sea.

The crews of the zeppelins must have learned that their craft had initially caused much terror among the islanders and took pains to show that they were friendly. Margaret Phillip remembers a zeppelin settling low enough over Queen's Park to drop a bag of mail for the American Consulate, causing quite a lot of excitement. Cecil Edwards remembers that the zeppelins would hover over the yard of the Great House at Gibbs' Estate at Moliniere where he played as a child. The crew of the zeppelin would exchange waves with the children. The Wildmans owned Sugar Loaf Island off Grenada's north coast. Once a zeppelin dropped some American apples for them while they were holidaying there. On several occasions, a zeppelin passed over the Church of England Primary School in Grenville and dropped tinned food and sweets in the yard for the children and teachers. This was much appreciated because these were really trying times for some people in Grenville. However, the first sight of a zeppelin always triggered a slight fear response.

IV: MOTOR TORPEDO BOAT FLOTILLAS

Early in 1942, England had sent the 19th Motor Torpedo Boat Flotilla to Trinidad to strengthen the defences of the Gulf of Paria and carry out limited escorting of inter-island vessels. MTBs were fast boats, equipped with armed torpedo tubes on deck and four depth charges. The following year, these MTBs were replaced by Flotillas. The 30th ML Flotilla comprised twelve Fairmile D motor launches. These launches were 115 feet long, carried heavy armament and twelve depth charges. They had a speed of 20 knots, and were able to hunt submarines and escort inter-island vessels. The other Flotilla — the 118th ML Flotilla —

comprised smaller Harbour Defence motor launches. Both flotillas were manned by a mixture of RN and TRNVR officers and men. Kelshall states that towards the end of the war, these launches were commanded mostly by Trinidadians.

V: TELECOMMUNICATIONS

Another defensive measure instituted in the Caribbean after 1942 was the vast network of radio stations and teletype facilities that was established even in the smallest inhabited islands. It was possible for a submarine sighting to be reported immediately, eliciting rapid response. A chain of radar stations was also established on the north coast of Trinidad that gave coverage from the Venezuelan coast to Grenada.

VI: THE U-BOAT MENACE LESSENS

The adequate anti-submarine defences established in the Caribbean had, by the end of June, 1943, caused a drop in U-boat activity. Several U-boats had been sunk, including two U-tankers. The loss of the tankers, on which the submarines were dependent for Caribbean operations, put to a halt the expansion of this activity. However, the U-boat Caribbean Operation had actually been designed to end. It was never conceived as a long-term tactic, but as fast, furious, and deadly. This it certainly was.

Although Germany lost the war, the operations in the Caribbean were highly successful. Of the 97 German U-boats and 6 Italian submarines[123] deployed in the Caribbean from February 1942 until April 1945, only 17 were sunk. This constituted two percent of the 784 U-boats lost worldwide. But for each U-boat sunk in the Caribbean, the Allies lost 23.5 merchant ships, making the U-boat offensive the most cost

---

123 Italian submarines operated near the Windward Islands.

effective campaign fought by Germany in any World War II theatre of war. The final figures of losses of shipping in the Caribbean Theatre were 400 merchant ships, with another 56 damaged. The losses in the Caribbean accounted for 15 ½ percent of the total merchant ships lost in the war.

Besides the anti-submarine defences in operation, an important factor leading to the demise of many U-boats was the Allies' acquisition of the means to crack Germany's highly secret Enigma code. The ability to know the positions of the U-boats meant that the U-boats could be hunted, spotted and destroyed. The U-boats now had difficulty in even achieving a safe passage from Europe to the Caribbean. The acquisition of this code, so important to the safety of the Caribbean, took place far away near the Nile Delta. A Sunderland Flying boat spotted U559 and transmitted the information to destroyers in the area. They converged on the U-boat and, faced with certain destruction, the U-boat's captain, rather than let the submarine fall into the hands of the British with all its secrets, gave the order to scuttle the vessel. The *HMS Petard*, one of the destroyers, lowered a lifeboat, and the U-Boat's crew swam toward it.

While the rescue of the German sailors was in progress, two British crewmen and a young trainee dived into the water. The crewmen entered the rapidly sinking vessel while the young trainee remained on the conning tower. The crewmen recovered the Enigma coding machine and codebooks — an invaluable prize for the British — and managed to hand them to the young man before they were pulled under and drowned as the submarine sank. The Germans never knew that the British had the coding equipment and could now crack their coded messages, and they could only wonder how the British were suddenly able to pinpoint the position of every U-boat and German vessel, and sink them with impunity.

By 1943, there were also regular Allied combined air and sea patrols in the Caribbean and in American coastal areas using the British armed motor launches, zeppelins, and submarine spotter aircraft equipped with anti-submarine weaponry.

Kelshall lists four other events that assisted the Allies in gaining control in the Caribbean theatre of war. One was Mexico's declaration of war against Germany in June 1942, after which Mexico opened up her air bases to US aircraft, allowing for greater coverage of the Gulf of Mexico and the Yucatan Passage. The second, following two months later, was the opening of hostilities between Germany and Brazil. The third event was the establishment of several Allied radio direction-finding stations on the US coast and in the Caribbean. The radio traffic between the U-boats and their headquarters in Germany gave the Allies the opportunity to triangulate the sources of radio traffic and to pinpoint the location of the U-boats, how many were operating, and to guide the submarine hunters to find and destroy them. The fourth event was the combining of the Leigh Light carried by RAF costal command aircraft with airborne ASV radar. The Leigh Light was a powerful 22 million candela carbon arc searchlight of 24-inch diameter, fitted to the RAF's Coastal Command patrol bombers and used from June 1942 onwards to spot surfaced U-boats at night. The aircraft would approach the submarine using its ASV (Air to Surface Vessel) radar and only switch on the searchlight beam during the final approach. The U-boat would then not have sufficient time to dive, and the bomb aimer would have a clear view of the target. The Leigh Light was so successful that for a time, the U-boats were forced to switch to daytime battery charging when they could at least see aircraft approaching. The ASV radar also allowed the Allies to spot the U-boats as they approached the Caribbean from their home base in Lorient.

The pilots of commercial aircraft also helped to spot the U-boats. On one occasion, a pilot on an aircraft belonging to the British West Indian Airways [124] sighted two U-boats surfacing in the Caribbean. A radio report warned nearby Allied forces and both submarines were sunk.[125]

By the middle of 1943, when the losses of the U-boats in the Caribbean reached the crisis point, Admiral Dönitz began to recall what remained of the Caribbean fleet. However, there was to be one last spectacular battle on 7th August, 1943 between a US aircraft and a U-boat. This took place off the north coat of Trinidad near Margarita Island between U-boat 615 under the captaincy of Ralph Kapitsky and U.S. aircraft from the bases in Trinidad. The battle lasted from 5 p.m. on 5th August until 4.50 a.m. on 8th August, 1943. This U-boat had stayed behind to act as a rear guard to allow the other U-boats in the Caribbean to depart.

---

124 This airline was started on 27th November, 1940, with flights from Trinidad to Barbados and Tobago. The aircraft was a Lockheed Hudson registered VP-TAE. Farfan lists the first pilot as "Snark" Wilson. The inaugural flight to Grenada was in January 1943.

125 Hitchins, P. 37

When it was spotted and engaged by US Navy planes, U615 fought so brilliantly and inflicted so much damage that the US forces refused to believe that only one submarine was involved. In the process, the U-boat absorbed enormous punishment from depth charges and strafing. When the position was hopeless, the U-boat was scuttled and 43 submariners were taken as prisoners of war, but Captain Ralph Kapitsky was injured during the battle, and instructed that he was to be propped up on the periscope standard so he could continue to direct the operation. He died at this station as he slowly bled to death. Kelshall writes:

> When the Hamburg built boat finally shipped below the waves, she carried the body of her gallant commander with her. [126]

One member of the crew of U615 also died in this battle.

The withdrawal of most of the U-boats from the Caribbean area resulted in the Caribbean Basin being downgraded by the US Military from D (exposed to major attack) to B (possibility of only minor attack). However, in spite of the recall of most of the U-boats, the U-boat Command needed to keep an appearance of activity in the Caribbean to pin down Allied resources to the area. A few more U-boats were dispatched to the Caribbean. One of the last U-boats to serve in the Caribbean was U530, captained by Lange, who returned to Lorient in August 1944. The last event regarding U-boats in the Caribbean was the arrival near Trinidad of twenty-two survivors from the *Oklahoma* sunk on 25th March, 1945, one thousand miles east of Trinidad. They were sighted and rescued eighteen days later near the Dragon's Mouth, Trinidad. This caused great consternation as it was believed that the U-boats were bottled up in the far north, and that the Caribbean and South Atlantic were completely safe. However, the new style XXI U-boats, including U532 captained by Fregattenkapitän Ottoheinrich Junker, were still prowling, and the sinking of the *Oklahoma* was his handiwork. Two U-boats were still at sea, but not near the Caribbean, when Germany surrendered.

The danger in the Caribbean was now deemed to have passed, and those who had known its seriousness and its extent relaxed. The Americans started to close some of its operations in Trinidad, although they intended to try to hold on to the base at Chaguaramas for the full extent

---

126   Kelshall, P 399

of its 99-year lease. All that remained now was the end to the war in Europe which was imminent.

Caution and worry over safety in Grenada were almost completely dissipated, and travel by sea from Grenada to other islands was no longer a cause for anxiety. Grenadians were now prepared to relax some of the internal controls that they had acquired and governed personal behaviour during the war. Evidence of this relaxation was that Grenadians would contemplate a sea-journey for pleasure, or allow their children to travel. There was little to cast a pall of worry in the minds of relatives, or to dim the holiday atmosphere of the 100 persons who boarded the two sturdy wooden schooners bound from Grenada for St. Vincent on 5th August, 1944.

PART V: THE STRANGE AND TRAGIC LOSS OF THE GRENADIAN AUXILIARY SCHOONER, THE ISLAND QUEEN

Painting by Susan Mains

I: GORDON CAMPBELL'S EXCURSION

Traditionally, the August or Emancipation weekend in Grenada lasted four days and was celebrated with beach picnics and horse-racing, both of which were popular among all sectors of Grenadian society. St. Vincent offered a more varied programme of events, full of activities including an annual gymkhana, numerous aquatic events, a hike to the crater of the Soufriere volcano, football and cricket matches, and other sporting and fun activities. Wartime had not broken the tradition.

There was, therefore, much interest when six weeks before the Emancipation/August weekend of 1944, an excursion to St. Vincent was advertised by Gordon Campbell, well known as an organiser of dances and concerts.

The proximity of Grenada and St. Vincent encouraged constant interaction between the populations of these two islands, with frequent travel between the islands by schooner and sloop. Although the war did not stop the travel, there was a falling off of inter-island competitions during the war, such as sporting events between Grenada and the other Windward Islands. One such inter-island competition, the Windward Islands Schools Football Competition had started just before the war. It was discontinued during the war and re-instated in 1947 along with tennis tournaments. These sporting events promoted healthy rivalry, lasting friendships, and even intermarriage between people of the two islands. The war was now drawing to a close, and the opportunity offered by Campbell to meet and socialise with the sports people of the neighbouring isle and to travel in the happy company of other young Grenadians was too good to miss. Grenadians who had family ties in St. Vincent, or who had family members in the Government service stationed in St. Vincent, also welcomed Campbell's excursion as an opportunity to pay visits.

Campbell planned that the excursion would leave Grenada on Saturday, 5th August, and return to Grenada on Wednesday, 9th August, early in the morning. Travel would be by the *Island Queen*, a schooner owned and captained by Chykra Salhab [127] of St. George's. The tickets for this excursion went like "hot bread" and Campbell, realising that the *Island Queen* could not accommodate everyone who wanted to go, decided to also hire the *Providence Mark*. Those who booked later got tickets for the *Providence Mark*, but all the young people wanted to travel by the *Island Queen*. They said it was because they wanted to experience the boat's powerful engine, but the primary factor governing their choice was that all the young people wanted to be in company with one another, and to party all the way to St. Vincent.

The news of the excursion spread to St. Vincent, and was looked forward to on that side as well. The excursion was mentioned in *The Times* newspaper of St. Vincent of 5th August, 1944. The newspaper said that

> A party of excursionists is due to arrive here tomorrow from Grenada to spend the August Bank Holiday. It is hoped that quite a lot will take advantage of the holiday to attend the Gymkhana (sic) Sports.

---

127 Chykra was, and still is, referred to by his first name, and so I will also call him "Chykra".

Painting by Susan Mains

II: WHO WERE THE PASSENGERS?

In 1944, Grand Anse, L'Ance aux Épines, St. Paul's and La Borie were still considered to be country. Westerhall and Fort Jeudy were undeveloped and considered "bush". Most of the population of St. George's lived within the city limits, some living quite respectably upstairs a shop or business place. The following streets were heavily residential: the Carenage, Tyrrel (now H.A. Blaize), Green, Lucas, Woolwich, Scott, Church, Young, Halifax, Hillsborough, Grenville, Granby, St. John's, St. Juille's, Bruce, Melville, and Cross. Nearly all the passengers who signed up for the excursion to St. Vincent, regardless of age, knew each other and lived on one or other of the streets in St. George's. There was hardly a stranger aboard the *Island Queen*.

The excursion comprised several overlapping groups, based on the reasons why they had signed up for the excursion. The majority of those who signed up for Campbell's excursion were young holidaymakers full of the joy of life, intent on enjoying what St. Vincent had to offer for the Emancipation weekend. Aboard the *Island Queen* were some of the most popular young people in Grenada, including some of the most beautiful girls and some of the most promising and good-looking young men, all looking forward to having a good time. This first group included: Lucy DeRiggs, a stunningly attractive girl and her handsome older brother, Hyacinth; Hans, Louise and Sheila Marryshow, beloved children of Grenada's outstanding politician and *Renaissance Man* T.A. Marryshow, and Laynika Scoon, a shy and quiet beauty. Reginald St. Bernard and his sisters Jocelyn St. Bernard and Lucy Paterson, all from Sauteurs, were also on board, as were Perle "Jappy" Moore Phillip and Doreen Scoon, vivacious and popular girls from Grenville.

Sylvia and Rita Gomas and their four other sisters were very musical and already in high demand to play for functions around town. Clythe, Erma, Ena and Eileen who were not going on the trip, went down with them to the boats to see Sylvie and Rita off. The two sisters were booked on the *Providence Mark*, and expected to have their cousin, Jack Baptiste, as a travelling companion. As they boarded, the young crowd on the *Island Queen* hailed Sylvia and Rita to come over to their boat and give them some music. This they did, along with Lennox

Howell, a good friend of Jack's. Persuaded by friends on the *Providence Mark* to stay on that boat, Jack did just this, first sitting down in the edge of the wharf with Lennox Howell to divide the food that they had packed together. The mental snapshot of the two lads at this task is the last memory of Lennox Howell that some people had, and which they treasured for the rest of their lives.

A second group travelling were the All Blacks Football Club that was made up of young men, now mostly civil servants, who were graduates of the Grenada Boys Secondary School (GBSS) and who were friends from boyhood days. Many of these would later enter the ranks of the professional class. The Club also included some non-playing members. The Club as a whole had decided to go on the August Holiday excursion to St. Vincent. A few weeks prior to the 5th August departure date, when it was clear that two boats would have to be used to accommodate everyone, Campbell asked the leader of the All Blacks Club if all of its members would travel on the "string boat". Even though they were some of the first to sign up for the excursion, Cosmo St. Bernard writes that:

> We did not have any objection to being relegated to travelling on the second string boat, the *Providence Mark*. [128]

And so it was that all the members of this club, except Huille Husbands, who wanted to stay on the *Island Queen* with his girlfriend Thelma Archer, gave up their places on the *Island Queen* and accepted alternative places on the *Providence Mark*.

A third group of passengers were invitees to the wedding in St. Vincent of Kathleen "Kypsie" Cruickshank, which was due to take place on Monday, 7th August, 1944. Kathleen's two sisters, Frieda Archer and Clarice Hughes, were from St. Vincent but had married into well-known families in Grenada. Much of Kypsie's wedding party was on board the *Island Queen*. Apart from Kypsie's sisters, Kypsie's eldest niece, Thelma Archer, was on board. She was about 20 years old and was to be the chief bridesmaid. Frieda Archer, Kypsie's sister and the mother of Shirley, Thelma, Ian and Terrence was a "fantastic needlewoman" and had made the bride's and bridesmaid's dresses, as well as outfits for herself and others in the wedding party. Another daughter, Shirley

---

[128] Cosmo St. Bernard, "The *Island Queen* Disaster." See Appendix III

Archer [129], who was then 13 ½ years old, had gone to St. Vincent a week before by plane carrying the dresses for the bride and bridesmaids. Frieda's other children, Ian, then 14, and Terrance, then 10, were left at home with their father, DeVere Archer, who was the headmaster of the Grenada Boys Secondary School.

Dawn Hughes, the daughter of Clarice and Russell Hughes, and therefore another of the bride's nieces, was to travel with the party going on the *Island Queen* but her father felt that she was too young to travel by schooner and arranged for Dawn and her aunt, Frieda, to travel by plane a few days before, taking up the rest of the outfits for the bridal party. When the time came, Frieda had not finished her own dress and decided to stay behind to complete it, and to travel later with the rest of the family on the *Island Queen*. Dawn went alone by plane. Tina Glasgow, a Vincentian, who worked for the Archers and was going to help with the wedding, was also included in the family group on the *Island Queen*. The wedding party's luggage aboard the *Island Queen* included a great big wedding cake, wedding finery and jewellery, wedding decorations and wedding presents both from themselves and from friends in Grenada who could not attend the wedding.

Apart from the relatives of the bride, there were other invited guests on board the *Island Queen*. Claire Patterson, who was to sing at the wedding, was aboard. A handsome couple aboard the *Island Queen* were Ian Hughes (the oldest son of Earle Hughes[130]), a very prominent young man, and Helena "Honey" Rapier, his pretty girlfriend. Honey was given that nickname because she was "as sweet as honey" and because of her beautiful golden-brown complexion. Ian had just successfully begged Derek "Doc" Renwick to exchange tickets with him so that he could be with Honey on *Island Queen*. Derek very generously allowed his friend to have his ticket and travelled instead on the *Providence Mark*.

Ian Hughes, Huille Husbands, Ivor Knight and Ena Fletcher, all fellow passengers on the *Island Queen*, were four of the six young people recently interviewed by the Medical Advisor to the Stockdale Commission, Sir Rupert Briercliffe, for medical scholarships from

---

129 Mrs. Eric Kelsick

130 Earle Hughes was a member of the Executive Council, and served on a number of Boards including the Board of Directors of George F. Huggins, the Grenada Co-operative Bank, the Bus Company and the Grenada Building and Loan Association.

the Colonial Office [131]. They were awaiting word as to the success of their applications.

A fourth group on the *Island Queen* was made up of very young passengers – school children – mostly from the St. Joseph's Convent, going home for the holidays, some taking their Grenadian friends with them. In 1944, St. Vincent had two high schools both run by the Government – the St. Vincent Grammar School for Boys and the Girls High School. However, St. Joseph's Convent and the GBSS were regarded by many outside of Grenada as surpassing any other schools in the Windward Islands for excellence. Wealthy persons in the Windward Islands, who wanted a good education for their children without sending them out of the region, would send the girls to St. Joseph's Convent and the boys to GBSS. The children would either board at the school, or with relatives or friends of their parents in Grenada. Two of these schoolgirls from St. Vincent attending the St. Joseph's Convent were Patsy and Jean Fraser, aged 14 and 16. They were the youngest of five daughters of Hon. Alexander "Alex" M. Fraser, a member of the Legislative Council of St. Vincent [132]. All his girls had boarded at St. Joseph's Convent in St. George's, where they had been outstanding pupils.

The Frasers usually travelled by air. Arrangements had been made for them to travel home to St. Vincent with a school friend from France for the holidays. However, their air passages were cancelled, as often happened in war time. Hearing about the planned *Island Queen* excursion to St. Vincent, Louie, the eldest Fraser sister, now married and living in St. Vincent's capital, Kingstown, with her father's approval, took the opportunity of booking the three children on the *Island Queen*.

---

131 The other two persons interviewed were Hilda Gibbs, who in later life became Grenada's first female Governor, Dame Hilda Bynoe, and Hyacinth Sylvester

132 The father of the girls, Alexander (Alex) Fraser, was born in St. Vincent Oct.14th, 1871. Tragically, on May 2nd, 1902, when Alex was 31, both his father and stepmother were killed from the emission of volcanic gas during the eruption of the Soufrière volcano in St. Vincent. Alex Fraser owned Rutland Vale Estate, where he raised livestock, grew sugarcane, cotton and arrowroot. He was a great friend of Fr. Charles Verbeke, a Roman Catholic Priest and architect, who had built both the original building of the Presentation Brothers' College in St. George's, Grenada and the Roman Catholic Church in Kingstown, St. Vincent. Alex Fraser discussed the matter of his daughters' education with Fr. Verbeke, musing that he was thinking of sending them to Codrington High School in Barbados. Fr.Verbeke convinced Alex that it would be far better to send them to the St. Joseph's Convent in Grenada. Fr. Verbeke personally introduced him to the nuns, and used his influence to persuade the Convent to take these Presbyterian girls into a school that was almost exclusively for Roman Catholics.

He gave his permission for the change in travels, particularly as he knew the captain of the *Island Queen*. However, the parents of their friend did not like the idea of their daughter travelling by boat and cancelled her holiday to St. Vincent. Instead, she returned home to Martinique.

Louie Fraser was a great friend of Marjorie MacLeish from the time they were both pupils at St. Joseph's Convent. Marjorie had spent holidays in St. Vincent with the Frasers and the two families had grown quite close. The day of the excursion, Jean and Patsy turned up at the MacLeish home to tell her goodbye before going to the pier to board the *Island Queen*. Marjorie MacLeish remembered the scene for the rest of her life. Jean and Patsy were wearing dresses and their school "Panama" hats, with a hatband marked "SJC". When the children left, Marjorie stood at a window overlooking St. George's harbour. The girls were clearly visible to her as they boarded the boat at the pier. She waved a last goodbye.

Merle Minors, Eileen James and Oona de Freitas were also pupils of St. Joseph's Convent who were going home to St. Vincent on the *Island Queen*. The final two school girls were pupils of the Church of England High School for Girls. These were Dawn and Denise Slinger, the daughters of Dr. Evelyn Slinger, the Grenadian surgeon at the Colony Hospital. At the urging of his Vincentian wife, Ena,[133] Dr. Slinger had agreed to allow his two girls to travel to St. Vincent on the *Island Queen* in the care of an older lady to visit their Vincentian grandmother, and to attend the Cruickshank wedding. They would stay with Adrian Date, the Acting Administrator in St. Vincent, whose daughters, Josephine and Alison, were of similar ages as Dawn and Denise. Denise had hoped that her friend, Norma Pilgrim,[134] would come with them to St. Vincent to spend the holidays. However, Norma's mother, Evelyn Pilgrim, said that it was not possible to send her on that excursion, but promised that she could go the following year. As a result, Denise wanted to remain in Grenada; nonetheless arrangements were made for both Slinger children to travel.

Shortly before the excursion, Dr. Slinger brought his daughters to see Eileen Gentle,[135] who had just had her first baby at the Private Block

---

133 née Richards
134 Mrs. "Eddie" Sinclair.
135 Née McIntyre

of the Colony Hospital. In her memoirs, she writes that on the day of the excursion she spotted the children on board the *Island Queen* as the boat passed directly in front of the Private Block. Boats exiting St. George's Harbour used to pass so close to the hospital buildings that anyone looking out for a boat can easily pick out individuals on deck. The children also saw her and waved to her. [136] The children probably also waved goodbye to their mother, who lived at Rathdune, the surgeons' residence situated on the hill near to the hospital, overlooking the sea. The *Island Queen* would have passed directly below the windows of this house.

On the day of the excursion, Chykra insisted that all the children on the excursion were to travel on the *Island Queen*. A family man himself, he preferred to keep them on his boat and under his supervision, even if this meant that they would be on the boat most likely to have passengers in very high spirits, who undoubtedly would have a party going before the boat cleared the harbour. This is why the school children were aboard the *Island Queen* among the partymakers, and not on the boat that would have the quieter voyage.

In addition to the many young people going on the excursion simply to enjoy themselves, the members of the All Blacks Football Club, also intent on enjoying the excursion and events in St. Vincent, the guests going to Kypsie Cruickshank's wedding, and children heading home to St. Vincent for the school holidays, there was a fifth group of passengers who were simply taking advantage of the excursion as a means of getting to St. Vincent. Two of these were Joan and Lester Richards, residents of Trinidad who were going to St. Vincent to see Joan's sister and her new baby. They were so anxious to get to St. Vincent that they arranged their entire passage by schooner, travelling from Trinidad and onward to St. Vincent on the *Island Queen*. Humphrey Parris, Chykra's brother-in-law, was in great need of a machine part for the sugar factory at the Grand Anse Estate, where he worked. He took the opportunity of the excursion to go to St. Vincent to look for the part, taking with him Donald Link, a young trainee at the factory.

Bernice Swapp was originally booked to travel on the *Providence Mark* but asked to change her ticket to go on the *Island Queen* to be with

---

136 Gentle, P. 93

friends. Ruby DeDier, one of her friends, recalls that Bernice slipped on the pier as she was about to board. Onlookers called out to tell her that slipping before boarding a boat was bad luck, and that she should go on the *Providence Mark*. She turned around to do just this, but others started to laugh at her, and call her superstitious, and so she turned back again and boarded the *Island Queen*. Aboard the *Island Queen*, also, was a man whose wife did not want him to go on the excursion and could not persuade him to stay at home, so she hid his money. After a bad quarrel with his wife, he frantically searched and found his money at the last moment, arriving at the pier in good time to board the *Island Queen*. [137]

Finally, Chykra had invited a few friends to go on the excursion as his guests. Only one, Joseph King, was able to accept his invitation.

III: WHO WERE THE CAPTAINS?

Chykra Salhab, a Grenadian born in Syria, was the owner and captain of the *Island Queen*. Now in his 60s, Chykra had come to the West Indies with his parents at the age of four, and by 1944, had been living in the West Indies for 46 years and in Grenada for 37. Until he married, Chykra lived in the Salhab household, a three-story house, which still stands, half-way down Scott Street. This household was presided over by Chykra's mother Dora, or "Mooma" as she was affectionately called. Chykra had two brothers and four sisters. All of the members of the Salhab family were well-known and well-liked.

Chykra was a skilful, experienced sea captain. On 11th July, 1939, while travelling to St. George's from Carriacou on *Cassandra*, another of his boats, his alertness caused him to spot and rescue three fishermen from certain death by drowning. For this action he was officially thanked by the Government of Grenada. He had been described as a "swashbuckling entrepreneur", and he and his two brothers engaged in many different enterprises. Under that exterior, however, Chyrka was a man of quality. He loved the game of Contract Bridge, and regularly played with H.H.

---

137 Redhead, P. 17

Pilgrim. He also played at the Diamond Club in Belmont owned by Hesketh Shillingford. Besides captaining the *Island Queen*, Chykra liked to sail for pleasure, often with Colin McIntyre. He was a generous family man who loved children. He had empathy and sympathy with friends and acquaintances and, like the rest of his family, would seek to ease a situation of hardship if he could.

When World War II broke out, the Government of French-mandated Syria aligned itself with the pro-Nazi Government of France. Chykra was still a national of Syria, and under the Defence Regulations, he was technically an enemy. Moreover, Section 9 of the Defence Ordinance stated that "Nothing in the Act shall qualify an alien to become the owner of a British ship." Chykra Salhab was sorely distressed over his predicament of nationality. He was very unfortunate because he had applied for British citizenship before war had been declared, and approval was pending. Grenada was the only country he knew and he was very well-respected, including by the Government. *The West Indian* newspaper of 9th June, 1940, issued an editorial in support of Chykra, stating that he

> Lived as a law abiding citizen and has been engaged in various undertakings of a character wholly unprejudicial to British interests.... (T)he greater part of his lifetime may be said to have been spent in the interest of Grenada's development in his particular sphere of activity. He has been a pioneer in the local motor traffic business; as a proprietor and employer of labour Grenadians alone have stood to benefit; as a health resort owner he has assisted the tourist industry and last but not least, as a motor launch and auxiliary schooner owner, he has helped in the development of inter-island trade, and personal relations. For over twenty years he has held Government contracts for various public services.

In the same issue of *The West Indian*, Chykra publicly stated that:

> I am willing to swear allegiance now to the British Crown as I regard myself as having owed my loyalty all along to no other power. In fact, I applied some months ago for citizenship papers. I know no one in Syria, nor anything about the country. The West Indies has become my home.

The West Indian newspaper mooted that Mr. Salhab's several friends would deeply appreciate the Government's leniency in first serving him with the notice according to the Defence Regulations, and then granting him temporary permission to continue captaincy of his boat, and to ply his regular route from Grenada to Port-of-Spain, Trinidad. This is exactly what happened, for Chykra was able to maintain his regular schedule to Trinidad, and moreover was at the helm of the *Island Queen* on 5th August, 1944, bound for St. Vincent.

The *Providence Mark* was owned and captained by Mark Hall, who in 1944 was in his early 40s. Hall was originally from Sauteurs, but had settled in Grenville. He had worked for some time in Alabama in the southern United States and had experienced the extreme racism practised there. This had left him sensitive to circumstances of racial prejudice. He was an experienced and even heroic seaman – one of the best.

Mark Hall is described by his sister-in-law, Drucilla Slinger, as a very dark, short, strapping, stocky, muscular man. He was helpful and kind, gruff but gentle, with a pleasant face. On his boats, Hall always wore "captain's whites", including a captain's cap. He would exchange this for a golfing cap when wearing ordinary clothes. He loved children and let them play around the *Providence Mark* when it was being built in Grenville.

Painting by Susan Mains

IV: THE SCHOONERS

The *Island Queen* was built on the Ballast Ground, St. George's, which is the piece of land now occupied by Camper and Nicholson's marina. She was a gaff-rigged auxiliary schooner, with graceful lines, and at 64 tons was heavier and larger than most Grenadian schooners. Designed by Ralph Ogilvie Sr. and built by local boat builders, the *Island Queen* was built using greenheart wood wherever this was suitable. She had a "canoe stern" without much overhang, and a curved bow without a bowsprit. The hull was painted black and the railings in gold.

To the youngsters who lived in St. George's and loved the sea and all on it, the construction of the *Island Queen* was fascinating and exciting. The construction site attracted almost daily visits, and the progress on the boat could be keenly observed. George "Porgie" and Julian Rapier, Ian and Dudley Hughes, and Gordon Preudhomme as boys had all been frequent visitors to the building site, travelling across to the Ballast Ground in the sturdy boats that they had made themselves out of scrap lumber. The launching of the *Island Queen* took place with great ceremony on Thursday 29th June, 1939. This was the feast of St. Peter, a day annually celebrated in Grenada as "Fisherman's Birthday". The launching enhanced the jubilation and festivities of this traditional feast day.

A detailed report of the launching appeared in the 2nd July issue of *The West Indian* newspaper. Some of those listed as present to witness the launch read like a Grenada *Who's Who:* Administrator William Leslie Heape and his wife, the Colonial Treasurer, I.C. Beaubrun and Mrs. Beaubrun, Senior Medial Officer, Dr. Edgar Cochrane and Mrs. Cochrane, Grenada's famous barrister and civic activist, C.F.P. "Jab Neg" Renwick and Mrs. Renwick, and the Honourables Arnold Williamson, T. A. Marryshow, and John F. Fleming. Other guests included additional high officials of Government, members of the clergy, businessmen, planters, and prominent citizens. Schoolchildren and a host of other Grenadians were also there as onlookers. The event was also witnessed from the hills that surround St. George's. Judith Parke saw the launching from where she lived at Richmond Hill.

In his address[138], Administrator Heape said that the honour of participating in the launching fell to him and his wife because Governor Popham and his wife were in England. However, he was happy to be officiating along with his wife, because

> All things related to the sea, to ships and the people who build and man them are very close to our hearts. My wife and I have watched her planks being laid and have seen her in various stages of construction. All here today can see that this is a fine ship and her lines a beauty to behold...
>
> Mr. Salhab, you have not only built a beautiful ship, but a powerful ship. May she be a fast ship, a dry ship, one that goes well to the windward. I appeal to you that, despite the engine you will be fitting in her shortly, you will not spare her canvas. Let her have lofty spars and canvas in full measure and let her show her pace as a sailing vessel. She is a great credit to the men who built her, and may she live long to be a great credit to the craftsmen of Grenada.

---

138 This was carried in its entirety in the *West Indian Newspaper* of 2nd July 1939.

Heape regretted that the boat's designer, Ralph Ogilvie, had died recently, and missed the boat's launching.

Boat-launching in Grenada follows strict protocol, one aspect of which dictates that the name of a boat is not revealed until the moment of "christening". At this moment, the name is whispered by the owner into the ear of the person "doing the honours" of christening the ship with a bottle of champagne. In this case, it was Mrs. Heape who had the privilege to receive the name of the boat from Chykra and declare: "I name this ship *Island Queen*. May she be a safe ship".

The West Indian newspaper recalls the amusement at the arrangements to ensure that the breaking of the traditional bottle of champagne on the ship's hull was "without risk of injury or (clothes) being soiled". Perhaps the organisers were too careful. On the first attempt, Mrs. Heape let the bottle fly, but the bottle hit the side of the ship with a dull thud without breaking. On the second attempt, the cork, which had already been eased, flew out as the bottle struck the side of the ship, spraying the ship with champagne, but the bottle remained still unbroken and half full, dangling from its line off the side of the ship. However, the champagne sprayed on the ship was deemed to be enough for a "christening", and in keeping with another boat-launching tradition, a white pennant bearing the name of the ship in red letters was then unfurled in the bow. To complete the ceremonies, the vessel was sprinkled with Holy Water and blessed by Fr. Alexis O' Brien O.P. of the Roman Catholic Church. Champagne was then "broken out" so that the guests could toast the vessel.

At 1.10 p.m. the job of moving the vessel into the sea began. Many lent a hand to roll the *Island Queen* into the sea. The first move, although only a few feet, was greeted with loud cheering by those present. However, floating a vessel is a slow process, and at sundown *Island Queen*'s bow had still not touched the water. The next day the work was completed, and the *Island Queen* "rode the waves" just after midday. On deck were Chykra's brother André and his brother-in-law Osborne Steele (married to Chykra's sister, Isabelle, called "Zabelle"). Osborne was also Chykra's very good friend, associate, co-shipbuilder and the mechanic for the *Island Queen*, who often travelled as crew to be able to attend to the engine, if necessary.

Sometime after its launching, the *Island Queen* was fitted with a powerful 125 horsepower German-made Krupps diesel engine. Willie

Redhead records that the engine for the *Island Queen* had been ordered from Germany and shipped via Trinidad.[139] Cuthbert Woodruff, a young customs officer at that time, who would eventually become the Anglican Archbishop of the Windward Islands, remembers being the officer to clear this engine. Unfortunately, the large, powerful German engine of *Island Queen* caused problems from the start. Firstly, the engine had to be started with compressed air, unlike the engines generally used for similar vessels. This would be a problem if the engine shut down in the middle of a voyage. Secondly, the propeller shaft proved too big and too long, thereby calling for alterations to the propeller sleeve. As a result, *Island Queen* began to leak, and there was "endless bother".[140] Finally it was found that the engine was really too powerful for the boat. On its maiden voyage, the engine almost opened up the hull of the boat, necessitating repairs. The engine could never be run at full throttle. However, these problems were solved, and Chykra was meticulous in the maintenance of his boat and its troublesome engine.

Most people, however, were quite unaware of these problems. The *Island Queen* was possibly the fastest boat travelling between Grenada and Trinidad at this time. She is remembered as being glorious to behold, especially when coming into St. George's Harbour in full sail. The *Island Queen* was used for general transport of goods and passengers to Trinidad and St. Vincent. During the war the role of the *Island Queen* and of other schooners and sloops became more important in transporting those seeking work in Trinidad, and the urgent need to bring back essential imported goods, foodstuffs, gasoline and kerosene.

The *Providence Mark* was a newer, but smaller, two-masted auxiliary schooner, built in St. Andrew's near Telescope Point in an area known as "The Pool". She was completed and registered in June 1944 and weighed 39 tons. Arthur Bain recalls that she was fitted with a 66 horsepower Kelvin engine, which was a very good diesel engine that was easy to start. He remembers also the *Providence Mark*'s beautiful lines, which were patterned after the famous Canadian Grand Banks fisherman, the *Blue Nose*, built in 1921. *Blue Nose* was a champion among her peers. The *Providence Mark* was faster than the *Island Queen* under sail alone, but could not quite keep up with her when she used both engine and sail. Both the *Island Queen* and the *Providence Mark* required a crew of four to five men.

---

139   Redhead, P. 15
140   Redhead, P. 15

Painting by Susan Mains

V: THE VOYAGE TO ST. VINCENT

In 1944, the major business places on the Carenage each owned a pier in front of their premises. The firm of Jonas Brown and Hubbard was the agent for the *Island Queen*, and therefore it was Hubbard's Pier and its precincts that exuded holiday atmosphere, albeit with some confusion on Saturday afternoon, 5th August, 1944.

Many of the young people who could not go on the excursion themselves, either came down to the pier to see their friends off, or watched from the windows of the houses on the streets that rise in rows from the Carenage, like an amphitheatre. Among the watchers were Marcella Lashley and Shirley Buckmire, who watched from the Jacob's house on Lucas Street. Excitement was high in anticipation of the departure of the two boats that would accommodate the people for Gordon Campbell's excursion. Gordon Campbell was feeling very pleased because this excursion was by far the most successful venture he had ever managed.

The time of departure had been set for 1 p.m. but the boats, which were ready, were unable to leave because there were too many people on board the *Island Queen*. For most young people, the *Island Queen* was definitely the boat on which to be. There would be guitar music, singing and merriment during the entire journey to St. Vincent. In fact, the party on the *Island Queen* had begun from about noon. Volunteers were called for to switch from the *Island Queen* to the *Providence Mark*, but those who didn't care which boat they travelled on had already exchanged their tickets in response to the pleading of the younger people.

Boats leaving Grenada for another island had to be cleared by both Immigration and Customs. The young customs officer, whose duty it was to see that the boats carried the permitted number of persons and no more, was supported in his task by Campbell and Captain Chykra Salhab. However, their efforts to persuade people booked to travel on the *Providence Mark* to take their places on that vessel were in vain. The passengers affected weren't even listening to the customs officer because they were having such a good time and were determined not to be left out of the continuing fun. Finding he was getting no co-operation at all from those already partying on the *Island Queen*, the customs officer requested that the boats move from the Hubbard's pier to the pier opposite the Fire and Police Station, so that a more senior officer could support him in reducing the number of people on the *Island Queen*. The boats were moved, and when it was absolutely clear that neither boat would be allowed to sail with the *Island Queen* overloaded, those holding tickets for the *Providence Mark* very unwillingly – some acrimoniously – transferred to the *Providence Mark*.

The clamour to transfer to the *Island Queen* by those booked on the *Providence Mark* was upsetting to Mark Hall, who felt slighted. He began to feel that his boat was deemed as "second class". This was not the case, as both boats carried persons from a cross-section of the society, but survivors of the excursion clearly remember the discomfiture the young passengers caused to this upright, good man by their thoughtlessness and their selfish and unruly behaviour. Willie Redhead recalls:

> One young man was bitterly disappointed and, in fact, demanded to know why his name was not among those on the "*Island Queen*'s" list. He was firmly told that it was not there, and that he should travel with the "*Providence Mark*". He kicked up a hell of a row, but nevertheless decided to transfer himself to the "*Providence Mark*" and sulked the whole voyage through. [141]

This young man would live to become a well-known and successful businessman and the father of several successful sons.

It took several hours for the passenger load to be satisfactorily settled, at least insofar as the authorities and the captains were concerned. The *Island Queen* and the *Providence Mark* finally prepared for departure, and their engines were again started sometime between five and six o'clock on that afternoon. Ernest Williamson had just returned home from the USA where he had pursued a course in motor mechanics. He had been working very hard in his father's garage, cleaning up the place, because the mechanic shop would now be placed under his jurisdiction, while his brother George took over the accounts. Both brothers were on the *Providence Mark*. When Ernest realised that the *Island Queen* was leaving with all the gaiety and merriment, and he would be left to have a very quiet voyage on the *Providence Mark*, he, on impulse, jumped from the *Providence Mark* to *Island Queen* while the *Island Queen* was already moving, almost falling into the water. His luggage

---

141 Redhead, P. 16

remained with his brother George on the *Providence Mark*. Because of the exchanges and constant movement of people between the boats even, as in this case, up to the last minute, the lists of names on the ships' manifests were no longer accurate.

The *Island Queen* eased away from the wharf, engine astern, turned into the stream and out of the harbour, "full of merriment – twanging guitars, the youthful voices lifting in gay songs of a tomorrow that would never be." [142]

As the boats departed, a young woman stood on the wharf quite bereft. She was supposed to have travelled, had gone to do something else during the long wait, and returned too late. She stood there weeping at what she believed was her misfortune. "Mike" Arthur had also missed the boarding on the Carenage. He was so determined to make the excursion that, instead of standing crying like the young lady, he had someone drive him to Fontenoy, where he got a fisherman to row him out to join the *Island Queen* as it came along.

As was customary, the *Island Queen's* sail would be hoisted when she cleared the promontory of Hospital Point, on which were located Fort George and the Colony Hospital. When the *Island Queen* was glimpsed by those who happened to look down Market Hill or similar vantage point on the other side of town, she was in full sail. She looked truly majestic as she headed for St. Vincent as the dusk drew in. Joyce MacLeod [143] from Mount Rose, St. Patrick's was a young nursing probationer at the Colony Hospital on 5th August, 1944. She was on the ward on the hospital's top storey when she was sent with a few other probationers to empty bedpans at the end of the building. The veranda at that end overlooked the Esplanade, which is the local name for Melville Street with its sea wall. She remembers it was very breezy, and that they saw people on the Esplanade waving gaily at a schooner going by. It was the *Island Queen*. One of the girls sighed and said, "I wish I was on that boat!"

Margaret Phillip, Monica Charles and Joy Cave saw the *Island Queen* departing from Margaret's bedroom window on Upper Grenville Street. The children waved at the boat, but they were reprimanded by Mrs. Phillip who made them stop immediately, because she said that waving at a departing boat was bad luck.

---

142 Redhead, P. 17

143 Mrs John DaBreo

The Invader was a sailing boat fitted with an engine whose proud owner was Colin "Skipper" McIntyre. Since he could not go on the excursion, Colin's plan was to follow the *Island Queen* up the west coast with a group of friends. The party on board included Ermintrude Hagley,[144] Alma Murray,[145] Porgie Rapier, Claude Patterson and Carlos Rodriquez. Porgie was given the helm, so Colin could keep an eye on the engine. Just before turning his boat opposite Beauséjour to return to St. George's, Colin drew close to the *Island Queen*, and *The Invader* and the *Island Queen* travelled parallel to each other for some time, with the passengers of both boats calling to each other and enjoying the sport. Porgie Rapier remembers seeing his friend Ian Hughes standing with Honey Rapier by the afterstays of the boat. He also remembers that Ivor "Punky" Knight was holding the headstay.

When it was decided that *The Invader* should turn back, in order to give the guests on board an easier time of it, she slowed to let the *Island Queen* draw away, and then turned gradually for home. Just as the boats started to separate, Porgie remembers shouting to Ian and Honey to "Have a good time". For many others aboard *The Invader*, the last glimpse of their friends on board the *Island Queen* just opposite Beauséjour would be imprinted on their memories forever.

The voyage to St. Vincent usually lasted between 12–15 hours. Having delayed departure almost six hours, the boats were now scheduled to arrive in St. Vincent at approximately six in the morning. One of the little excitements of travelling in tandem with another boat was to see which boat could reach its destination first. The *Providence Mark* departed about half an hour after the *Island Queen*, under engine and sail, but those on the *Providence Mark* still hoped that she would hold her own and beat the *Island Queen* to St. Vincent.

The *Island Queen* and *Providence Mark* stayed on a parallel course for the full length of Grenada's western coastline. The *Island Queen* positioned herself further out to sea than the *Providence Mark*, while the *Providence Mark* hugged the coastline about five miles offshore. Even after dark, the boats could see each other's lights, which were mounted on the masts. Cosmo St. Bernard remembers seeing the light of the *Island Queen* between 8–8.30 p.m., almost on the horizon. This

---

144 Later the wife of the first President of Trinidad and Tobago, Sir Ellis Clarke

145 Mrs. Paul Kent

was when the *Providence Mark* was close to Duquesne Bay. By the time the *Providence Mark* cleared the north point of Grenada, those travelling on her were delighted to observe that *Providence Mark* had overtaken *Island Queen*. The *Island Queen* was then was about 10–12 miles to the west. Alister Hughes's account of the event also puts the *Island Queen* behind the *Providence Mark* at about 10 o'clock that night [146]. At that time, Hughes says, both vessels were to the lee of Carriacou, and the lights of the *Island Queen* were still visible, astern and to the west. This account is corroborated by that given in *The West Indian* of 10th August, 1944, which says that by midnight, sight was lost of the *Island Queen*, "though a speck seen in the distance could have been her sails, but of this there could be no certainty." If the speck was indeed the sails of the *Island Queen*, this was the last known time that Grenadian eyes had glimpsed her. The *Island Queen* separated from the *Providence Mark* going further and further away from the island chain, choosing the "steamer route" that would take her further still out to sea, with the object of travelling a shorter triangular route to St. Vincent.

Inter-island schooners usually docked in the sheltered area of the Kingstown Harbour, near where the Cruise Liner Tourist Facility is today. When the *Providence Mark* docked about 9 a.m. on Sunday, 6th August, the passengers were slightly surprised, but delighted that they had arrived before the *Island Queen*. After being cleared by customs, most of the passengers waited around for their friends on *Island Queen*, fully expecting that she could not be far behind. Shirley Archer remembers going down to meet the boat around 9 a.m. with Dawn Hughes, who awaited her mother, her cousin, her aunt, and Tina. When the *Island Queen* failed to arrive after a reasonable time, they had to leave. They were told that the *Providence Mark* had passed the *Island Queen* at Carriacou at about midnight and that it was just delayed somewhere, somehow.

---

146 Hughes, P.45 The Official Report gives the time of 10.30 p.m. when the boats lost sight of each other.

Painting by Susan Mains

VI: WHERE IS THE ISLAND QUEEN?

By 10.30 a.m., when the *Island Queen* had not yet reached St. Vincent, it was clear that she was not just lagging behind the *Providence Mark*. Something must have happened to delay her arrival. The matter was reported to the Police and the Harbour Master, and telephone calls to Grenada and Carriacou were made to see if *Island Queen* had had engine trouble and had either returned to Grenada or put in at Carriacou. The news that came back was that the *Island Queen* was in neither place.

The Acting Administrator, Adrian Date, [147] and his whole household were generally and particularly concerned as they awaited the arrival of the Slinger children, who were to be their little guests. However, because everyone believed that the *Island Queen* had gone adrift, a major panic was delayed. If the engine had shut down, it could not easily be restarted because of the need for compressed air to do this. If the *Island Queen* was travelling under sail alone, then it could have been that she was blown off course, or was drifting with the current if the winds were not favourable. Some excursionists and those waiting for the *Island Queen* made light of the non-appearance, saying that the *Queen* must be down in Carúpano again. *Island Queen* had drifted down to Carúpano on another occasion when its engines had failed and there had been no wind [148]. Moreover, Edgar Adams observes that it was not entirely unheard of that ships would arrive several hours later than the time expected [149].

Nevertheless, by midday, the matter was worrying enough for the Grenadian authorities to call those in charge of the British Fleet Air Arm in Trinidad asking whether any distress signals had been picked up or if anything unusual had been seen. Air and sea searches by the Fleet Air Arm and the Hudson motor launches based in the Grenadines

---

[147] Adrian Date demitted the office of Acting Administrator as planned, a few days later, on 14th August, 1944.

[148] Redhead, P. 16. The same fate was suffered on another occasion by the *Principal S*. Among the passengers on that trip was Reginald Palmer, who was knighted by the Queen later in life, and who served as one of Grenada's Governors-General. Sir Reginald was returning to Teachers College in Trinidad.

[149] Adams (2002), P. 142

were requested at the same time. However, nothing unusual had been reported; the sea had been calm and the weather fine.

The passengers on board the *Providence Mark* who were to go to the Gymkhana or to other events went to those activities in the expectation that the *Island Queen* would soon turn up, but the continued non-appearance of the *Island Queen* threw a damper on the holiday outing. The report in *The West Indian* for 10th August said that in fact "Kingstown's air of festivity was distinctly spoilt". Anxiety got progressively worse as the hours passed without news.

Louie Fraser had married George Blencowe, a Vincentian, in 1941. The young couple made their home at Windsor House overlooking the Kingstown Harbour. Arrangements had been made to have Jean and Patsy, her young sisters, walk up the hill to refresh themselves before travelling to their parents at Rutland Vale in the country. When her sisters failed to arrive, Louie sent a message to her parents suggesting that they come to stay with her while they anxiously awaited news of the boat.

Tragically, all who kept watch for the arrival of the *Island Queen* watched in vain for a boat that would never dock.

VII: THE WEDDING

Despite the worry and growing anxiety, it was decided that the wedding of Kathleen Cruickshank would take place as planned on Monday, 7th August, 1944, as the expectation was that the *Island Queen* would turn up somewhere in the next few days. It was a pity that the bride's sisters, her niece, Thelma, who was to be the chief bridesmaid, and many friends who were on the *Island Queen*, would miss the wedding. A good friend of the bride, Lenore, was the same size as Thelma and could wear the chief bridesmaid's dress. She was Thelma's stand-in. Dawn Hughes cried and cried when her mother did not arrive in time for the wedding. This was only the prelude of her grief.

On the afternoon of the wedding, there was a storm brewing, and some wedding guests recall that as the bride walked down the isle of the church, there was lightning and a peal of thunder. Many remember having "chills run up their spine" at this. Everyone had "disaster" tucked into the back of their minds as far as the *Island Queen* was concerned, and the thunder just seemed like a bad omen. However, at the reception, some guests decided to push the questions hanging over the fate of the missing into the back of their minds. Shirley Archer recalls that many wedding guests "drowned their sorrows", others "had a ball," while still others remained glum and close to tears.

Newspaper reports of the wedding were carried in both St. Vincent newspapers at the end of the week. The cloud of anxiety overshadowing the occasion was not mentioned in either article. *The Vincentian* of Saturday, 12th August, entitled its story "Wedding Bells." It was reported that

> On Monday afternoon there was a pretty little wedding at the St. George's Cathedral, when Kathleen Ianthe, daughter of Mr. and Mrs. O.C. Cruickshank [150] of Kingstown was given in matrimony to Mr. George Donald Sherman.
>
> The ceremony was fully choral, the Revd. Woodruff presiding at the organ. The ceremony was presided over by Revd. Fr. Ivo Keown-Boyd. Miss Lenore Marshall was the Chief Bridesmaid, the other maids being Misses Shirley Archer and Dawn Hughes, nieces of the Bride. Mr. A.V. Sprott performed the duties of best man. The Philharmonic Orchestra, of which the groom is conductor, was in attendance...

After the ceremony the bridal party met at the home of the bride's parents for the reception, where a number of friends were entertained.

VIII: THE SEARCH BEGINS

When it was learned that the *Island Queen* had not arrived in St. Vincent, some Grenadians set out in their own boats to look for her, but returned having seen only the empty sea. On Monday, 7th August, Gittens Knight, the Acting Colonial Treasurer, decided that an official

---

150 In later life Kathleen Cruickshank ran her parents' guesthouse in Kingstown, until she migrated to Canada.

search from Grenada should be undertaken. He contacted Telfer Preudhomme, the owner of the schooner *Rose Marie* to arrange for this boat to search along the route of the *Island Queen*.

The *Rose Marie* was a two-masted schooner, fitted with 44 horsepower Kevin engine. She had been converted to a "Q" boat by the U.S. Navy, and was fitted with hidden armour piercing artillery capable of disabling a U-boat. She was "on call" to the U.S. Navy to be chartered by them to search for U-boats in the waters near Grenville and along the east coast of Grenada. When the *Rose Marie* went on these missions, she remained under the captaincy of Sydney Wells, [151] but he was under the supervision of American personnel and the boat had a crew of Americans. The *Rose Marie* never had her capabilities as a "Q" boat put to the test. The usual crew of the *Rose Marie* was on holiday, but Captain Sydney Wells immediately began to make enquiries as to whom he could get to assist him in their absence. George BOW Otway had gone hiking in the forest of his estate, Beaton Place, St. David's, and a messenger was dispatched to find and fetch him to help man the *Rose Marie*. Mr. Otway arrived in St. George's at 11 a.m. ready to go. Captain H.E.A. Rowley[152], who was in Grenada on holiday, was approached to join the others on the *Rose Marie*. This he was glad to do. In the meanwhile, the *Rose Marie* was equipped with food, fuel and ballast and awaited instructions to depart.

Administrator Heape had joined Gittens Knight at Grenada's Security Office. The Royal Navy authorities in Trinidad were contacted again by cable and asked for an urgent updated reconnaissance report. This message was repeated to Barbados, St. Lucia and St. Vincent. There was an immediate response, and Grenadians saw the reconnaissance plane pass over St. George's at 3.30 p.m. Not too long after, however, the report was received that nothing had been seen of the *Island Queen*. The *Rose Marie*, with Captains Wells and Rowley and Mr. Otway on board, was then instructed to depart, and left at 6.30 p.m.

---

151 Sydney Lawrence Wells was one of the best-known boat captains in Grenada, known throughout the length and breadth of the island. He was of average height, slim and wiry, and "West Indian White". He was uncle of Arthur Bain, Brenda Williams and Hermione Greasley. A veteran of World War I, he had also worked in the oil fields of Venezuela. He lived in Grenville, but moved around Grenada, pursuing his various interests. He captained the *Rose Marie* for her owner, Telfer Preudhomme.

152 Captain Rowley was the brother of the distinguished Grenadian social worker, Pansy Rowley.

The instruction from Gittens Knight to Captain Wells was to heave off Gouyave. If Knight received any further information, a Revenue Officer would row out to the *Rose Marie* to give details. There was no further news awaiting the *Rose Marie* at Gouyave, and she proceeded towards Carriacou, where again there was no news. At midday on Tuesday, 8th August, the *Rose Marie* was instructed to return to St. George's. This she did, still maintaining a search by following a different south-westerly course, returning to Grenada at 9 p.m. In the meantime, Gittens Knight stayed at the pier, questioning the captain of every boat that came in, but no one had seen any sign of the *Island Queen*.

When the *Rose Marie* returned with no news, the Administrator sent a further lengthy cable to the British authorities at Trinidad, stressing that this was now an emergency and asking for further help from the Fleet Air Arm and the United States Navy. This cable was again repeated to St. Vincent, St. Lucia, Barbados and in addition it was sent to Venezuela, Curaçao and British Guiana. Assurances were received that air searches would be initiated in the likely areas of the *Island Queen*'s drift.

Up to and including Tuesday, 8th August, it was still presumed that the *Island Queen* had only gone adrift, but concerns now grew incrementally as the hours passed. The Executive Council of the Government met in emergency session several times in quick succession, and they did all they could, given their limited resources, to find the *Island Queen*. On 8th August, Gittens Knight again dispatched a lengthy cable to the Naval Authorities in Trinidad, requesting that the sea and air searches be extended to include the entire area of sea between the island chain and meridian 65° west, as far north as Dominica and as far south as the Venezuelan coast. This was agreed, and in addition zeppelins also joined the search.

Painting by Susan Mains

IX: A SAD HOMECOMING

Most of the passengers on the excursion who travelled on the *Providence Mark* left St. Vincent for Grenada late on the evening of Tuesday the 8th August for Grenada. Margaret Bain [153] was fortunate to get a

---

153 Mrs. Michael Bain

passage on the *Providence Mark* for the return trip from St. Vincent to Grenada. She was living in St. Vincent with her family because her father was a magistrate stationed there. Her family had been invited to the Cruickshank wedding but she could not attend because she had an awful toothache. Her father went looking for the first passage on a reliable boat to send her to Grenada for dental treatment. Even though she was in terrible pain, she remembers the passage well. When she boarded the boat, she noticed Shirley Archer, Dawn Hughes, Ewart Hughes, and Margaret "Buddy" Pantin on board. She relates that everyone on board firmly believed that they would see the *Island Queen* docked on the Carenage when they arrived in St. George's. But alas...

When the *Providence Mark* docked at 6 a.m. on Wednesday, 9th August, in St. George's, a large crowd was there to meet her. As soon as she moored at the Hubbard's pier, the captain was besieged with questions from anxious and sombre relatives and friends, some of whom were unsure as to who had travelled on which boat due to the last minute "swaps". Some of the returning young people went into the joyous embrace of parents and spouses who thought they had travelled on the *Island Queen*.

Everyone wanted to know what the captain thought had happened to the *Island Queen*. When did he lose sight of her, and weren't they sailing together? Skipper Hall replied:

> I never saw her again once night fell. She was way out to sea and travelling much faster than I was. Yes, the weather was O.K. No squalls or anything to worry about. [154]

X: EVERY EFFORT MADE

Later on Wednesday, 9th August, the Executive Council in Grenada met again in emergency session for a review of what steps had been taken to discover the location of the boat. The meeting was addressed by Captains Sydney Wells, H.E.A. Rowley, and George Otway, who shared their thoughts and observations with the members. Now that the *Island*

---

154 Redhead, P. 17

Queen had been missing for four days, the question of food came up. The Executive Council was told that the *Island Queen* had on board sufficient food and water to last a week, if judiciously rationed. *The West Indian* of 10th August listed the food on board as "250 lbs of vegetables, half bag of flour, half bag of rice, a barrel of biscuits, peas," and other articles.

By 13th August, two British vessels, the security launch *Clipper* and the *El Altman*, had completed a search of the Grenadines and found nothing. On 14th August, the area of search was enlarged to include as far west as meridian 71° west longitude, as far north as Puerto Rico, and as far south as Venezuela. There was no point in searching to the east because the winds and currents made it impossible for a disabled or drifting vessel to be carried to the east.

Part of the search was observed by Brenda Wells. Brenda was spending her summer holidays at Prospect in St. Andrew's with Clarence and Sheila Gun-Monroe. On hearing the news that the *Island Queen* was missing, the Gun-Munroes took Brenda with them to visit May and Jack Copland at Mount Rodney, where they could observe some of the search activity. Brenda remembers seeing the planes passing overhead, going north, and vessels out to sea following the route of the *Island Queen*, searching but not finding the missing schooner. Two more weeks went by with absolutely no news.

Miss Estelle Garraway was a former headmistress at the Church of England High School for Girls. She is remembered as being a very sweet person, tall, thin, and unmarried. As a headmistress, she was said to be absolutely fair in her dealings with the girls. In her private life, her avocation was the operations of the psychic realm. Known as a visionary, psychic and clairvoyant, she indulged in reading tea-cups and foretelling the future for people. She was well-respected for this, because many Grenadians believed in the possibility of seeing into the future and treasured clairvoyants like Miss Garraway. At this time, Miss Garraway "saw" in a dream the *Island Queen* moored among the islands off the north coast of Venezuela. Encouraged by this dream, wealthier families with children on the *Island Queen* commissioned another search.

Since the area of search was to be the islands in the north of Venezuela, the cooperation of the Venezuelan authorities had to be obtained and the search made in a Venezuelan craft chartered for the purpose. Captains Wells and Rowley departed from Trinidad on 6th September as representatives of the families who had commissioned the search. In the Venezuelan vessel *Vera* they scoured "every scrap" of the north coast of Venezuela and the offshore islands to see if they could find wreckage or other signs of the *Island Queen*. They also questioned residents and fishermen of these islands. The search lasted eight days. Captains Rowley and Wells returned to Grenada on Thursday, 14th September. In spite of a diligent search, no sign of the ship or any of the passengers was found, and the people questioned had no information to give to those waiting in Grenada.

> So they turned away dejectedly and consigned their lot to God. They sought his help in prayers and supplications. So the days passed into weeks and the weeks into months... NOTHING! [155]

Although the question of the costs of the searches came secondary to the Government and people of Grenada, these still had to be met. The Government approved the payment of £100 sterling on 17th October, 1944, to the owners of the *Rose Marie* for the searches done on 7th and 8th August, plus an amount of £12.10s in compensation for fares lost as a result of her having to leave Grenada out of schedule in connection with the later search of the Venezuelan coast. It was also agreed that the Government should pay the cost of the search of the Venezuelan coast, which amounted to £203 sterling. Even though some of the families involved contributed, the feeling was that Government should meet the cost, and that it should also enquire whether the Government of St. Vincent might be willing to subscribe. In the minutes of a meeting of the Finance Committee of Grenada, held on Friday, 23rd February, 1945, it was noted that the Administration of St. Vincent agreed to pay approximately £342 sterling, which was about half of the cost of the searches for the *Island Queen*.

Those with loved ones on the *Island Queen* individually made every effort to seek information from any source. For example, nearly every day Madeline Parris, utterly bereft, visited the American Consulate in St. George's, seeking information and begging that the Consulate "Do something". But she could get no information, because there was possibly none to be had.

All through the time of searching and waiting, the people of Grenada and St. Vincent prayed for the discovery and safe return of the *Island*

---

[155] Redhead, P. 18

Queen and her passengers. Norma Pilgrim remembers constant prayers being offered at St. Joseph's Convent for the girls who went there. Mother Columban was the well-liked but strict headmistress of St. Joseph's Convent. Porgie Rapier describes her as a very likable, remarkable person. She was a "dear soul", who insisted on precision in language. She was very fond of Jean Fraser, who was very bright, had a pleasing disposition and a beautiful singing voice. Mother Columban uncharacteristically wept openly on several occasions when prayers were being said at school for the students at the convent who had been on the *Island Queen*. Carlyle John remembers the prayers said in the St. Patrick's Roman Catholic School and at church for those lost on the *Island Queen*, and especially for the Joslyn and Reginald St. Bernard and Lucy Patterson. Carol Bristol, originally from St. Lucia, had been sent by his parents to Grenada to attend GBSS. He remembers that every Sunday after the disappearance of the *Island Queen* prayers would be offered in church for those lost on the boat. Vincentian Bertram Arthur recalls nightly outdoor prayer meetings held in Kingstown, St. Vincent, to pray for the missing, conducted by a Grenadian called Telesford. The hymn "For Those at Peril on the Sea" was always sung.

Painting by Susan Mains

XI: WHAT THE NEWSPAPERS SAID

The newspapers in Grenada and St. Vincent played an important role in providing the facts concerning the excursion to St. Vincent. Written for the time, they became an invaluable historical source. At the time the *Island Queen* went missing, the newspapers kept the public abreast of search activities, other news, or confirmation of information already passing from person to person. The editorials on the missing boat and its passengers in the newspapers in Grenada and St, Vincent tried to calm fears, keep hope alive and also to deflect the criticism of the Government's handling of the situation.

The lists of the passengers of both *Island Queen* and the *Providence Mark* had previously been published in the papers but because of the movement of passengers from one boat to the other, these lists were no longer accurate. Nevertheless, people went back to the lists, for it was one thing to hear about passengers missing; it was another thing if you knew them personally. Some cut out and saved these lists, and there are many copies put safely away in the papers of those who were contemporaries of the passengers of the ill-fated August 1944 excursion to St. Vincent.

The West Indian newspaper for Thursday, 10th August, 1944, carried a report detailing what was known so far about the ill-fated excursion to St. Vincent, and relating all the measures that had been taken up to that point to trace the ship. The editorial expressed the view that

> For the present, however, we are disinclined to take any pessimistic view as to the safety of those on board as incidents of the kind have by no means been uncommon.

In addition, the editorial called for more safety measures for Grenadian boats:

> There are yet many features connected with the distressful occurrence which call for uncompromising investigation, especially in view of the recent history of wartime happenings in our waters and repeated references in these pages both editorially and by correspondents to the need for greater safeguards for those who sail from our ports.

However, the editorial asked that

> Every citizen do his or her part while the tension lasts by refraining from indulgence in pessimistic rumour, bearing in mind the many at home to be affected by such an attitude to the incident.

This editorial also praised the Administration of Grenada, and interestingly introduces the idea that the *Island Queen* might have been destroyed by a U-boat.

> Thankfully, the Administration had done all that is possible to ensure a search for the overdue ship. Well do we know that in periods of submarine activity in these parts, air patrols have not always met with immediate success in their searches for shipwrecked seamen.

Finally, the editorial addressed the discontent in the population regarding official action, defends the Government, and asked people to have

> Patience, courage and a firm trust in the almighty, who rules the elements to which the missing are now exposed.... Investigation and fault finding can only be based on facts we sincerely hope it will be possible to have revealed in the very near future.

On 12th August, *The Vincentian* newspaper carried the following announcement:

> The *Island Queen*, a schooner that left Grenada on Saturday evening last with a party of holidaymakers for St. Vincent has failed to arrive, and up to the present, nothing has been heard of it. We understand that there are quite a number of children and a few Vincentians on board.

As the days passed, with nothing seeming to be happening and no further news, frustration mounted and some people started to lash out and to apportion blame. Most of the criticism, however, was levelled at the Government for not doing enough to find the boat and its passengers. It was also believed that the Government knew more than it was divulging.

In reaction to accusations against the Government, Alfred C. Coard, a well-known civil servant of Green Street, St. George's, wrote a letter to the editor of *The West Indian* on 11th August. It was published in the issue of 13th August. Coard commended the Government for prompt action in requesting help from the naval authorities in Trinidad. The facts, he said

> (C)ondemn and disprove rumours of the past few days that the Government was either callous or indifferent in the matter where the lives of so many people (women and children especially) were involved. This is not the time for attaching blame or censuring anyone, (that time will come, and come it must). In the meantime each and every one should refrain from expressing views of the "judgment passing" nature as no good purpose will be served by so doing, but rather create an atmosphere of rancour and ill will and unnecessary bitterness to the sorrow and mental anguish now being endured by the families of the loved ones aboard.

The letter ended by saying that, instead of criticism, people should exert and focus all their efforts, prayers, and hopes to bring about a speedy and happy reuniting of the unfortunate passengers with their respective families.

The issue of *The West Indian* for the 13th August also carried an article captioned "*Island Queen*'s Party's Ordeal not Uncommon". The article began with a note of hope.

> Absence of news of the *Island Queen* up to last evening should not daunt the island-wide spirit of hope that the schooner's fifty-six passengers and ten crew are yet safe.

> As we have said earlier, other vessels have been missing for longer periods because wind, waves and tide are fickle elements, having counterparts in the air above....

This last statement was based on all too frequent facts. The same issue of the newspaper carried an account of the harrowing experiences of Justice G.E.F. Richards, a Puisine Judge for the Windward and Leeward Islands, and others travelling on an auxiliary sloop chartered to take them from St. Lucia and Dominica for a sitting of the Circuit Court. Leaving St. Lucia with a fellow passenger, Mr. V. Beaumington of Barbados, and five crew, the sloop put in at Martinique enroute and left there on 24th July to complete the journey. Unfortunately, the sloop was caught in a hurricane and was blown off course. After nine days, it reached Testigo on the eastern Venezuelan coast. The newspaper reported that the sloop had carried only one day's supply of food, and after that was finished, the seven on board survived on crates of mangoes and avocadoes that Justice Richards had been carrying to Dominica, and on one bird and two fish that the crew caught.

Isaac Chin, a Trinidadian writing in *The West Indian* of 19th August, related his own tale of survival. He and five friends had boarded the *Principal S* for a holiday in Grenada. The boat ran into foul weather, and the wind and waves sent the boat a considerable way off course. The *Principal S* took three days to reach Grenada – a voyage usually of only a few hours. Chin said that, since arriving in Grenada, he and his friends were deeply affected by the news of the *Island Queen*, but

> I am of the conviction that the missing party is either still afloat lost, or have landed somewhere about. I don't see why anyone should infer anything else but that.

On 14th August, the people of Grenada and St. Vincent were assured of the concern of the Governor through a communication issued by the Government. Appearing in both *The Times* and *The Vincentian* newspapers on 19th August, and signed by Gordon Mancini, the Governor's Secretary, it read:

> His Excellency the Governor has requested the Administrator to assure the general public through the medium of your Press, of his deep concern for the safety of the Schooner *"ISLAND QUEEN"*, and his continuous sympathy with all those connected with the Grenadian and Vincentian passengers on board from his first receipt of the news on 6th August.
>
> All possible steps are being taken with a view to locating the missing vessel, but so far unfortunately without success.

The West Indian for 15th August further reported the deep distress of the Governor of the Windward Islands, resident in St. Lucia. To counter the rumblings and rumours, the newspaper added that:

> It would serve no useful purpose for the Government to withhold any positive information, regardless of its nature, when any news of such an order is available.

The public was supported with other messages of care, concern, pleas for calmness and hope, and faith in God's mercy. A letter to the Editor of *The West Indian* carried in the edition for 16th August, reminded the public of previous breakdowns:

> Sailing masters with whom we have spoken of the *"Island Queen"* express surprise at the attitude of those who regard the vessel's overdue time of reporting as extraordinary. They tell of instances on end of sailing craft being missing for several weeks.
>
> The land lubber cannot easily comprehend the ways of a ship at sea. How much harder must be his striving to know the mysterious working of the Hand moving the waves to calm or turbulence! While search continues and reveals nothing untoward we would serve the distressed better to remember that there is wideness in the mercy of that same Hand far wider than the Caribbean Sea of immediate contemplation.

On 17th August, *The West Indian* newspaper observed that the *Island Queen* had now been missing for ten days, and therefore could be approaching the South American mainland, or an island near the mainland. Hope should be high of word reaching Grenada of the *Island Queen*'s whereabouts.

Slowly, however, a different note — one of surrender to the inevitable — began to creep into the newspaper reports. The editorial of *The Times* for 19th August noted:

> There have been heart-beatings, and throbbings, in many a home in Grenada today and for the past few days over the fate of a schooner which left with holiday makers for this island, a week ago and up to now nothing has been heard of its whereabouts. To make things doubly worse there are quite a few children aboard. It has been stated that the ship had only foodstuffs to last a day or two as would be obvious coming on such a short voyage...

The lessons that could be learned from the plight of the *Island Queen* and other inter-island schooners also form a part of the editorial.

> Quite a few vessels have been reported missing for the past few months; one or two have reached the Venezuelan coast while nothing has been heard of the others... Never before in the history of these islands has such a toll on shipping been made as during the past few months, and we should not allow such conditions to go on without limit.

The editor then made a call for the introduction of more reliable and faster craft for the transportation of passengers between the islands. The editorial of *The Vincentian* newspaper of 16th September, also called for some good to come out of the *Island Queen* tragedy. The editor called for all schooners to be fitted with "rockets" for use in an emergency.

The editor of *The West Indian* at this time was T.A. Marryshow, three of whose children were missing on the *Island Queen*. His editorials and news reports were as much to inform and console himself and his family as it was to reach out and do the same to other families.

XII: CLUTCHING AT STRAWS

In spite of the changing tone of the newspapers, hope that the *Island Queen* would suddenly appear was very much alive. Rumours were frequent, but hearts sank lower each time a rumour proved without substance. Every vessel that came into the harbours at Kingstown or St. George's were met by anxious relatives and friends, to see if they carried survivors or had news of *Island Queen*. Lincoln "Brim" St. Louis remembers:

> I was affected as a resident of the Carenage at the time. On several occasions, it was rumoured that the *Island Queen* had returned. On each occasion, a large crowd of which I was a member ran to the pier only to discover that it was untrue. I was, however, eager for information about the causes of the disappearance of the boat. I never heard an official reason, but I got lots of unofficial explanations for the disappearance from older and much older boys sitting under the lobby of the Empire [156].

---

156 The Empire was a popular cinema, "liming spot" and beloved landmark on the Carenage that was demolished in 2005 to make way for a business enterprise.

A whole vessel with a boatload of passengers could not just vanish between Grenada and St. Vincent without a trace. That was impossible. It had to be somewhere — it had to come back. People kept watching, waiting, and clutching at straws.

To add to the general anxiety, the *Providence Mark* left St. George's for Trinidad on Wednesday 16th August on her regular run with forty passengers. Overnight, she was caught in a hurricane while about 45 miles off Trinidad. When word reached Grenada that the vessel had not reached its destination, it was almost too much for the population. Fortunately, and to everyone's relief, the *Providence Mark* limped back into the port of St. George at 8 p.m. on Thursday, 17th August, having been blown off course by the winds, and ending up well to the north of Grenada. Captain Hall, ably assisted by friend and fellow captain, Captain John Scott of the *Principal S* and his crew, had saved the ship and all the passengers. The only casualties were his cargo of bananas, which he had to throw overboard to lighten the ship. Unanimous admiration for Captain Hall's courageous and skilful handling of the ship during the tempest was reported in *The West Indian* newspaper of 19th August. Paula Julien [157] knew people who had seen Captain Hall's hands all raw and bloody from having held the tiller against the raging wind and waves.

Around this same time, too, a boat was sighted coming into the St. George's Harbour with one of its masts broken and its sails in tatters. News spread that it was the *Island Queen*. Cries of "Hurrah! The *Island Queen* is coming" spread like wildfire, and people rushed down to see. Alas, it was a false alarm. This was another schooner that had been badly battered by the same hurricane. This damaged schooner was observed coming in by Judith Parke's grandmother who, from the vantage point of her home on Richmond Hill, had been searching the sea relentlessly since the disappearance of the *Island Queen*. She hoped to see the *Island Queen* returning, and now joyously thought the moment had come.

The Captain of the *Lady Angela*, which had arrived in Grenada on 20th August, reported that he had seen an unidentified vessel which could have been the *Island Queen* at anchor north of Margarita off Blanquilla Island. This was reported to the Senior British Naval Officer in Trinidad by the Government. On 21st August, an advisory was sent to

---

157 Annette and Paula Julien were daughters of Willan E. Julien, a prominent Grenadian. Annette married and became Mrs. John Smith, while Paula became Mrs. Sydney Williams.

the Venezuelan authorities asking for their co-operation in identifying the vessel, and should survivors land on their north coast. Requests were also sent to Venezuela, Curaçao and British Guiana, asking that these countries look out for the vessel, which could have drifted south. A special air search by the Fleet Air Arm was requested and carried out on 22nd August. A telegram was received on 23rd August from the Senior British Naval Officer in Trinidad and printed the next day in *The West Indian* that there was indeed a sailing vessel there, but it was not the *Island Queen*. This telegram also contained a statement of declining interest by the military in Trinidad in following up leads and rumours from the Grenada Government regarding the *Island Queen*. The telegram plainly stated that "additional naval surface craft for this work would be unavailable" for further requests for searches without additional information. This was because

> Every possible step to find *Island Queen* has been taken and surveillance of area continues. It must however be realised that in view of the time elapsed and very thorough search conducted the chance of getting news of vessel is now remote. Word would be quickly got if any survivors had landed on Margarita Island.
>
> As previously stated all adjacent shorelines have been and still are subjected to close scrutiny from air.

The indications were that there was nothing more that could be done. The Governor of the Windward Islands was also preparing the people for the worst. A message appearing in *The Vincentian* of Saturday, 26th August, read:

> Though I am far from having lost hope of the safety of *"Island Queen"*, I nevertheless cannot refrain from conveying now to all those who have relations and friends aboard, the assurance of my heartfelt and constant sympathy in hours of anguished suspense through which they are passing.

Painting by Susan Mains

XIII: OFFICIAL CLOSURE AND MESSAGES OF CONDOLENCE

Esmond Farfan, writing about the loss of comrades in the RAF during the war still raging in Europe, observes that

> When your son or brother or friend is listed as missing, hope is stronger than grief, and so, the immediate pain is dulled. The real grief comes when hope is dead. [158]

So it was with those who had sons, daughters, mothers, fathers and friends on the *Island Queen*. On 29th August, the Governor of the Windward Islands issued a message that indicated that people of the islands should no longer hope that the *Island Queen* would return. Reproduced in *The Times* of 2nd September it read:

> A close surveillance still continues but it must be realised that in view of the time which has elapsed and the very thorough search which has been conducted, the chance of getting news of the vessel is remote. It is with sorrow that I am now forced to the conclusion that the schooner *"Island Queen"* must now be presumed lost with all hands.

Governor Grimble went on to express his heartfelt sympathy to those in Grenada and St. Vincent who had relatives and friends on the *Island Queen*, and to thank all those who had assisted in trying to locate the missing vessel. On the same day, the Secretary of State for the Colonies sent this telegram:

> I have learnt with greatest regret news that the Schooner *"Island Queen"* must be presumed lost with all hands and desire that you will convey my sincere sympathy to the people of Grenada and St. Vincent who must now be presumed to have lost relatives and friends on this vessel.

In sending this message to the newspapers, Administrator Heape appended his own heartfelt sympathy to the next-of-kin in their grievous loss.

The Vincentian newspaper in its editorial for 2nd September noted that:

> The Government Communiqué issued this week with reference to the loss of the *"Island Queen"* has given shock to all classes in the islands of Grenada, St. Vincent and St. Lucia. The sympathy of everyone goes out to the parents and relatives of those unfortunate passengers.

And so now the heavens were stormed, with tearful prayers for those who most people now accepted were gone forever. Several requiem services were held in Grenada for the passengers on *Island Queen*.

---

158 Farfan, P.114

Margaret Phillip and Shirley Archer, both very young at the time, remember a very well-attended service being held in Grenada at the Cathedral of the Immaculate Conception, St. George's. This was a sung requiem mass held on 26th September at 6.30 a.m. with Rev. Fr. Denis Fitzgerald O.P., the parish priest, officiating. Margaret remembers another big memorial service at the Church of England for the travellers on the *Island Queen* held on the same day, when a service of the Holy Eucharist was held at 8.30 a.m. with Archdeacon Piggott as the chief celebrant. She recalls that the Church of England was overcrowded, and throughout the service there was uncontrolled crying and loud wailing. She remembers learning the hymn "O God, Our Help in Ages Past" to sing at the service. The singing of this hymn brought on increased expressions of grief. Alister McIntyre remembers these two services, and that there were services at the Scots Kirk and Methodist Church as well. The churches were all decorated with black, the services were extremely sombre and the sorrow absolute.

Smaller services and prayers were also held all over Grenada and abroad. There was a service held at the Church of England in St. Paul's for those lost. Annette Julien was a pupil at the Bishop Anstey High School in Trinidad in 1944. She remembers the whole school praying at a special assembly for Mrs. Joan Richards from St. Vincent, who was a teacher there, and one of the passengers of the *Island Queen*. A "Service of Sorrow" was held on 10th December at the Court Lily of the Valley in St. George's for the loss of D. Redvers Rowley.

Individuals sent up their unceasing individual prayers. The headmistress of the Church of England High School for Girls was Miss Mab Bertrand. She led the whole school in hoping that the pupils on the *Island Queen* would return. For that reason, the desks of the missing children were kept as they had been before the disaster for an entire term. Mrs. Slinger had large portraits of Dawn and Denise prepared and put on display for the assembly at the beginning of the new term. These portraits caused some grumbling, for there were at least seven past pupils who had been lost and other pupils had lost relatives on the *Island Queen*. Although these were not commemorated, neither had anyone come forward with a request for special remembrance on their behalf.

Messages of condolence and sympathy poured from all over the Caribbean to the newspapers on Grenada and St. Vincent, and directly to the bereaved. The editorial in *The Times* for the 2nd September, 1944, points out that many on board the *Island Queen* had been coming to visit loved ones from whom they were separated by a narrow strip of water. The editorial of this issue tried to convey the sense of bereavement and loss that spanned the twin colonies. In part it read:

The Work is O'er

The Cruel Fates have with one fell blow cut short those trees so rich in the blossoms of promise. The shock resulting has been sudden, and the blow severe....

No more stunning, no more horrifying, no more shocking, no more heart-rending information could have conveyed to any community than this. Men, women and children have perished in this terrible disaster, — a disaster which has no equal in living memory of these islands. Many a home today in our sister Colonies have been bereft of a loved one, it may be a father, a mother, a sister or a child. We can only sit and conjecture their sad fate, and the more we think, the more dismal the imagery appears in our minds – mothers and children clinging to each other as they literally bury themselves in the watery deep; here and there the anguish of a helpless one. What could be sadder than this?

Hope springs eternal, says Pope, but there are times and seasons when we must banish hope, when it is obvious that no matter what hoping we do it would all be of no effect. Each succeeding day since the ill-fated schooner left the shores of Grenada fresh hopes sprung within our breast; finally the end has come and all our hopes are thwarted....

It has been said that the suddenness of death takes away the consciousness of dying. If this is so in this case then there could have been no suffering, no pain, such as would be endured by those who die on their beds. Since there is not even a sign of wreckage, nor a dead body, it must be assumed that death was not only sudden but instantaneous. In this respect there is at least consolation to know that the writhings and agonies such as are experienced by sufferers ordinarily were entirely absent...

In (words of) the Funeral oration of Pericles uttered over two thousand years ago "They gave their bodies to the deep and received for each his own praise a memory that will never die, but shall grow deeper in the hearts of those who loved them...."

The West Indian for 31st October printed a message from His Grace the Archbishop of the West Indies, Dr. A.H. Antsley, that had been sent to the Bishop in St. Vincent. It read in part:

> May I sympathise with you most sincerely in the loss of the *Island Queen* which I see from today's paper must be taken as certain. It is a most distressing tragedy...Please convey to the sorrowing relatives my sincere condolence with the prayer that God will comfort them in their distress.

The Editor of *Onward*, the Diocesan magazine of the Church of England, also printed the Archbishop's message, and added his own that read in part:

> It is with a full heart that I write this message of sympathy to all who have lost friends and loved ones in the *"Island Queen"* disaster. Some of the missing are lives that could ill be spared and were held dear in the hearts of many. To my certain knowledge there were those who were trying always to live near to God. The mind goes back to happy days when some of them were younger and we held out for them high hopes of a valuable and useful future. Our consolation lies in the fact that our hopes are undiminished, even though they may not have the fulfilment we envisaged. God knows better than we what the future may hold, but of this we may be sure – He will keep them still in the folds of his great love and mercy....

One of the messages acknowledged in *The West Indian* of 28th October was from a former Grenada cricketer now living in Trinidad who particularly mourned the loss of Oswald Callendar. He wrote that Callendar "made himself very dear to me through his kindness in many directions". Other messages of sympathy printed in *The West Indian* were from Miss Clara Purcell, who was holidaying in Jamaica with her nephew, Justice J.E.D. Carberry, Mrs. Viola Wickham [159] and her family from Barbados, and Revd. S.W.C. Cross and family from British Guiana.

---

159 Viola Germaine Wickham was the widow of Barbadian journalist, Clennel Wickham (1895-1938), who had been given a job at *The West Indian* newspaper by his friend and colleague, T.A. Marryshow, in 1934. Described as an idealist, a champion of democracy and of the underprivileged, Wickham was forced to leave Barbados as he found that he could get no employment in his own country after daring to criticise the establishment. His *Herald* newspaper was closed and sold to settle a lawsuit that he had lost to a planter who claimed that he had defamed his character. Wickham died in Grenada of a ruptured appendix and is buried in the St. George's "Centre" Cemetery.

Neither the official nor unofficial messages brought complete closure. There was a minority who still did not give up hope for their lost loved ones on the *Island Queen*. In October, at the request of the Government of St. Vincent, another search was requested, this time to search Aves (also called Bird) Island, about 90 miles west of Dominica. However, the Administrator of St. Vincent telegraphed the Governor Grimble on 7th October suggesting that before a search party was dispatched from St. Vincent, air reconnaissance should be done, possibly from St. Lucia. The Governor replied on 13th October saying that the British Senior Naval Officer had informed him that United States Authorities had kept Aves Island under constant scrutiny during the period of the search for the *Island Queen*. No sign of any activity on Aves Island had been observed. By the time the search for the *Island Queen* was finally abandoned, 200,000 square miles of sea had been searched.

At that time, the populations of Grenada and St. Vincent were very much smaller and much more closely knit than they are today, the population of Grenada being 72,387 and that of St. Vincent being 61,647. In addition, the two islands were linked through intermarriage and friendship ties. In each island, also, there were deep ties of affection between people that transcended the barriers of the class structure. Therefore, the loss of people on the *Island Queen* was deeply mourned by the entire populations of Grenada and St. Vincent for many years.

The business of tying up matters regarding unfortunate passengers of the *Island Queen* was finished by the end of the year. The Government paid gratuities to the next of kin of all Government employees who had lost their lives. The amount of the gratuity was determined by the length of service and amount of dependents left behind. The gratuity in respect of D. Redvers Rowley, for example, was fixed at £248 sterling, equal to 90 day's pay. There is a sad little note relative to the widow of Redvers Rowley, Iris Rowley née Bertrand from Grenville. In order to collect the gratuities and ex gratia payments, the next-of-kin had to apply to the Court for Letters of Administration and become the legal representative of the deceased. The Executive Council minutes of 22nd May, 1945 noted that

> Owing to the circumstances attendant on her husband's disappearance, she (Mrs. Rowley) did not wish to apply to the Court for Letters of Administration in order that she

might claim the death gratuity... and asked that the gratuity be paid to her without the usual formulation.

Mrs. Rowley was advised that she had to conform strictly to the law. Although she could not face up to the finality of the death of her husband, deprived of her husband's income, and now financially strapped, she eventually went through the procedure to claim the gratuity.

The Colonial Secretary's end of the year Budget Speech included reflections on the grief and loss attendant on the loss of the *Island Queen*. The speech was reproduced in *The West Indian* of 13th December. In part the speech read:

> This has been one of the biggest tragedies in the history of the Colony – so many families have been affected; so many young people cut off in the flower of their lives. The Council, will, I am sure, wish me to take this opportunity of recording its deepest sympathy with all those both here and in St. Vincent who had relatives on board the vessel. Not only does our sympathy go out to them but also our great admiration of the fortitude with which they have borne their sorrow....

After the year's end, people tried to resume their normal routine but wondered if they would ever really be happy again.

Painting by Susan Mains

XIV: RACHEL WEEPS FOR HER CHILDREN

In Rama was there a voice heard
Lamentation, and weeping, and great mourning,
Rachael weeping for her children,
And would not be comforted
Because they are not.

Gospel of Matthew Ch. 2 v 18

The event, as would be expected, left permanent psychological scars on relatives and loved ones of the lost. The coping mechanisms used by both individuals and the society to counter crushing grief, and to try to make sense of something that was totally overwhelming in their reality, make an interesting psychological study.

One of the first reactions was denial. There were those who kept a constant vigil, expecting that their watch would be rewarded with the sight of the *Island Queen* coming back with their loved ones. Almost every afternoon after the *Island Queen* went missing, Mrs. Marie Scoon used to take her younger daughter Winifred to the sea wall of the Esplanade to see if she saw any sign of the vessel. It is said that Mrs. Scoon never gave up the hope of seeing Layinka again, always believing that the passengers of the *Island Queen* were perhaps on some deserted island somewhere. Iris Rowley also kept vigil for her husband Redvers Rowley. She would sit by the window looking out over the harbour in hope of seeing the *Island Queen* bringing her husband home. The pattern of watching for the boat that would never dock extended to the general population. People would go down to where vessels were tied up on the Carenage and "just stand".

Mrs. Slinger firmly believed that her daughters were coming back. When some of Dawn and Denise's classmates came to visit, she showed them her fridge stocked with goodies and treats for her daughters that they would have on their return. Dr. Evelyn Slinger, the children's father, was away from the island when his daughters left for St. Vincent, and when he heard the news he immediately returned to the island. To those who saw him during the next few weeks, it seemed as if he walked about in a trance.

"Brim" St. Louis remembers feeling very sorry for Louise Marryshow's mother, Edna Gittens, who was deemed to have been dealt one of the worst blows of the tragedy, having lost three children. Adina Maitland witnessed, with so many other residents of Tyrrel Street, "Auntie Edna" crying her heart out every day, sitting with her remaining children at her window, looking out to sea, hoping the boat would come back. The community of Green Street and Tyrrel Street ensured that she was visited daily during the first few weeks. The teachers of the Church of England High School for Girls, where her girls had gone to school, also came to offer her solace. After Edna regained some composure, she would dress nicely and visit Marcella Lashley's mother, Ethel Lashley, who lived nearby. The two friends used to sit together in Ethel's bedroom, which had a window looking out to sea, Edna always hoping she would witness the *Island Queen*'s return.

Some people's reaction to the disaster was to take refuge in fantasies. Some parents came up with amazing suppositions attempting to give assurances that their children were not dead, but taken, and were alive somewhere on the planet.

Of those who accepted the loss during the first few days, many claimed that they had had premonitions before the event. Austin Hughes, one of Earle Hughes' sons and Ian's brother, remembers that Pamela, their younger sister, begged Ian Hughes not to go, because she had dreamt that he had lost his life on this journey. But the pull of the excursion was too much to miss over a dream. "Zabelle" Steele, Chykra's sister, had a premonition, and begged Jean (Chykra's daughter) and his nieces, Elinor and Agnes, not to go on the excursion. They had been looking forward to the trip and visiting friends and family in St. Vincent, but listened to Zabelle's pleading and did not add to the monumental mourning in the Salhab household. Thelma Knight [160], 18 at the time, lived with her great aunt Marjorie Springer in a house right beside the lime factory on the Carenage. She asked her permission to go on the excursion, but her aunt said she was not to go, without giving any reason. Thelma implored but her aunt was adamant. Thelma had to be content to go down to the pier to see the *Island Queen* leave. Thelma firmly believes that this was the second of two separate occasions when her aunt, Marjorie Springer, had a premonition that saved her niece's life.

There were also a multitude of dreams after the boat went missing. Austin Hughes remembers that at 12.10 a.m., around the time when the *Island Queen* was supposed to have met her fate, his father was awakened suddenly by a gush of wind. There was Ian, his son, standing at the foot of his bed. Ian disappeared before he could ask him what he was doing there. Marie Scoon, Layinka's mother, dreamt that the boat had returned, and all was well. Oris Teka dreamt of Lucy DeRiggs, her friend, who appeared like a mermaid. Another person is reputed to have dreamt of a cross in the sky, and still another of a lot of dead people downstairs of a structure, all burnt beyond description, with the exception of Ian and Clarice Hughes. Shirley Archer and others remember that Miss Garraway dreamt of Chykra Salhab pleading for help. It was in response to another dream of Miss Garraway that the *Vera* had been commissioned to fruitlessly search the Venezuelan coastline for ten days.

---

160 Mrs. Thelma Phillip

About ten years after the disappearance of the *Island Queen*, Porgie Rapier awoke one night to a vision of his childhood friend, Ian Hughes, standing at the foot of his bed, looking as if he had aged ten years. Porgie was greatly surprised and shocked, and asked Ian: "What happened, man?" Ian replied to his friend: "They said I could come back to see you, but I must not say a word about what happened." Ian then disappeared.

The severe grief could cause anxiety, depression, collapse and even death. The reality, although accepted, proved too much for some of the older people, hastening their decline. Miss Molly Hill, the mother of Reginald and Joslyn St. Bernard and Lucy Paterson, lived on upper High Street, near McDonald College in Sauteurs. When she lost her children, she suffered from acute anxiety. Anastasia La Guerre remembers that at first she walked continuously up and down Main Street with her hand to her head in distress, hoping that she would meet someone who could tell her anything about her children. Soon after, she collapsed completely and had to take to her bed, repeating to herself the circumstances of her loss. While she was in this state, the community supported her with regular visits. Lynda Lalbeharrysingh went to see her several times with her mother and sister, because the mothers were friends. Eventually, Molly was able to get up and resume her life, but she was never the same again. Basil Bonaparte remembers that Molly Hill trained him for confirmation in the early 1950s. The children noticed her mood swings, and that she would sometimes stop teaching and lapse into a reverie. As young as they were, the children understood that Miss Hill was still in deep grief and had patience with their good but moody teacher. Basil thinks that she died shortly after he passed through her confirmation class, her lifetime cut short by her heartbreak.

Dora Salhab, Chykra's mother, was a second person who was never the same again. She cried for days, and walked in misery up and down Scott Street. She had not only lost her eldest son, but also her son-in-law, Humphrey Parris, and Thelma Steele, who was the sister of Osborne Steele, her daughter Zabelle's husband. Elinor, one of her grandchildren, had lived with Dora since a toddler. She was her constant companion and support. Unable to read, friends would drop in to read the Bible to her, which was a great consolation. Later Dora found consolation in baking and cooking, and baked for every visitor and almost the whole of Scott Street, hailing people from her window and sending food down to them. Although Dora eventually "caught" herself, she lived for only a few years more after the loss of the *Island Queen*.

The strain of the loss of his two youngest daughters proved too much for Hon. Alex Fraser. He fretted constantly, and was shortly after hospitalized with very high blood pressure. The newspaper report on his condition stated that his illness has been accelerated by anxiety for his daughters who had been aboard the *Island Queen*[161]. Fraser's episode of high blood pressure led to a stroke, and he was never able to speak again. He died on the 8th January, 1945, just five months after his daughters' disappearance. His wife now had her grief compounded. She went to stay with her daughter Louie and her husband at Windsor, St. Vincent, until the couple and their three-year old daughter immigrated to Canada. Then, another daughter Agnes, moved to Windsor from Barbados with her family. Louie remembers that until her death at age 93 in 1989, Mrs. Fraser spent long periods of time looking out of the window at Windsor above the Kingstown Harbour, being very quiet and crocheting as she gazed at the pier where the *Island Queen* should have docked decades before.

Louie Fraser Blencowe never gave up hope. She found it too difficult to accept the loss of her two sisters and many friends who travelled on the *Island Queen*. All her life, Louie clung to hope of finding her sisters or even some explanation of what happened to the *Island Queen* on the night of 5th August, 1944. One example of clinging to a faint hope was in October 1963, when a story was broadcast on CBC TV in Canada relating the story of two young white females in their twenties, who had been washed up on a beach in Africa during the war. The story stated that they were found barely alive by two nuns of a nearby convent and died shortly after. Louie immediately wondered whether by some miracle and against all odds [162] these two could have been survivors of the *Island Queen*, even her sisters, or at least persons who might have some information on what had happened to her sisters. She tried to get in touch with CBC to get more details, but all efforts to get more information from the television station failed.

Accepting the tragedy did not make reality easier to bear. Joan Bonaparte

---

161 This was first reported in the *Trinidad Guardian* and reproduced in *The West Indian* of 25th August, 1944.

162 Cf. footnote 181

remembers that on the day it was accepted that the passengers on the *Island Queen* would, most likely, never return, she witnessed a big crowd of people at the Gomas' house on St. John Street where she also lived. The crowd was so large that the veranda was filled, and people spilled out into the street, mourning and weeping for Sylvia and Rita, and talking about how excited these girls had been to be going on the trip, and how carefully they had selected the outfits that they were taking with them. Paula Julien remembers going with her father, Willan E. Julien, to pay a visit to Russell Hughes after the boat went missing. More than sixty years later, she vividly remembers how grey and "awful" his face looked, and his obvious grief and desolation at the loss of his wife, Clarice. The whole population of Grenada was glum and stunned, and St. George's was "very quiet." Anthony DeRiggs recalls that a "great sadness" overcame his grandmother, the mother of Lucy and Hyacinth DeRiggs, He goes on to say that quite commonly "pots were burnt while people peered into the Caribbean Sea."[163] People went into momentary sad daydreams, recalling, wondering, thinking, and grieving. Dunbar Steele remembers being depressed and frightened at the news of the loss of the *Island Queen*. He recalls the day that the loss was confirmed was "The saddest day in the whole of Grenada", when everybody seemed to be crying. Leo Cromwell, his face mirroring deep emotion over sixty years later, describes the weeks and months after the loss of the *Island Queen* as a "sad, sad time".

Sheila Marryshow had a small school in her mother's house situated on Briggs Alley that ran between Green and Lucas streets. "Brim" St. Louis, a past pupil of that school, was about 9 or 10 at the time. He remembers that there was "much song, dance and play activity" in the curriculum. It registered on his young mind that now he would be sent to another school, because Miss Marryshow was not coming back.

Unfortunately, some people resorted to anger as a coping mechanism, and indulged in the "blame game", searching for people to be made scapegoats for the tragedy; some of the things said were unreasonable, unkind, untrue and even ridiculous. Incredibly, among those who were blamed for the tragedy was Mrs. Archer, with whom fault was found for having her sister plan her wedding for St. Vincent and for encouraging the young people to go to the wedding! People also blamed Chykra, and

---

163   DeRiggs, P. 249

some followed the lead of the Official Enquiry which seemed to intimate that Chykra had been incompetent. This was very unfair because those who really knew Chykra were confident that if there had been anything this experienced and excellent seaman could have done to save the boat and bring back the passengers, he would have done it. There was talk of the *Island Queen* having been unsafe, forgetting that it was only once or twice that the *Island Queen* had had any sort of trouble with its engines, and it had travelled countless times and returned safely. The traditional informal racing contest between boats travelling together was condemned because it was said quite untruly that it caused the captains to take risks and compromise safety. This sort of talk deeply hurt the Salhab and Archer families who were already overburdened with grief.

A source of the anger was the feeling that the authorities were withholding information. The slightest departure from the norm became of greatest importance and was immediately attributed to the disappearance of the *Island Queen*. One such incident was the presence in Grenada of a British naval patrol boat shortly after the disappearance of the *Island Queen*. Instead of berthing alongside the pier as was customary, the boat anchored off Morne Panday. Her launches were observed running constantly to and from ship and shore. The rumour was that this patrol boat had brought the remnants of the victims from the *Island Queen* for identification.

> People flocked to vantage points to get a glimpse of this mysterious patrol boat carrying something. They forgot that a war was still on and all they could see was a grey hull in the dying evening, which soon lengthened into night ...By morning the ship was gone. [164]

Apart from the psychological trauma, there were also economic difficulties as a result of the non-return of the *Island Queen*. Several breadwinners were taken away from families, leaving them in desperate economic circumstances. Iris Rowley had to shoulder the financial responsibilities of the family, taking in boarders and sewing to support herself and her family. Madeline Parris and her children, Noreen, Joan and James (Jimmy), were allowed to live in the Grand Anse Estate Manager's house for one year after the *Island Queen* was lost. After that, she and her children had to give up their idyllic life on the estate

---

164   Redhead, P. 18

and return to her mother's crowded house on Scott Street. In order to bring in some money to help support her family, she opened the *Café de Paris* on the ground floor of the house, where she served lunches and later in the day provided refreshments for couples wanting somewhere to go on "dates". The café proved both popular and long-lived, for it was still there to provide a recreational spot for the Welsh Fusiliers who were stationed at Tanteen in 1951 during the demonstrations and disturbances that marked the rise to prominence and power of Eric Matthew Gairy.

T. A. Marryshow had been a great friend of Humphrey Parris, and Joan Parris remembers that he stayed in touch with the family and helped the family in many different ways, seeing that all the children finished secondary school. Madeline Parris never remarried. Her daughter, Joan, quotes her as saying that she "loved Humphrey so much that she would never, ever look at another man."

Septimus Alleyne, one of Chykra's crew, left behind a girlfriend, Alice Dumont, and two small children one between one and two years old, and a one-month old infant. Alice found herself sunk not only in grief, but in dire financial straits as well. She had to move out of the apartment that she and Septimus had shared and return to live with her parents. Fortunately, she was morally and financially supported throughout her ordeal by the Alleyne family.

Some children lost their mothers. Perle "Jappy" Moore Phillip left an infant son, Michael. Michael's father was at the time a Sergeant in the West Indian Regiment serving in Italy and Egypt. Ivan David, who was in the same company as Phillip and his fast friend, remembers the day the terrible telegram arrived for Phillip. They were sitting under a big tree talking when Phillip was handed the telegram which read simply:

> *Island Queen* missing ten days. Wife on board.

Phillip was devastated and chain-smoked for days. He was given compassionate leave, during which time he contacted his family in Grenada and made arrangements for his infant son.

When Justina Glasgow was asked to go with Mrs. Archer to St. Vincent, she left her daughter, Monica Charles, with her friend, Beatrice Phillip, as she usually did when she was going to be working late. Beatrice's daughter, Margaret Phillip, was around the same age as Justina's child, Monica, who was quite at home in this family. When Monica asked where her mother was, she was told that her mother had not returned from her trip. Eventually the child stopped asking. She stayed with the

Phillip family for about a year until she was sent for by her father who lived in Trinidad.

Imperceptibly people adjusted, and intensity of the grief faded into a dull ache, gloomy fatalism, painful remembrances, and "what ifs."

Painting by Susan Mains

XV: QUIRKS OF FATE

When the news of the excursion first began to circulate, there was hardly a young person in Grenada who did not wish they could be a part of it. Finances kept some people at home, but other quirks of fate came into play that helped determine who should be on the *Island Queen*, who should travel on the *Providence Mark*, and who should stay in Grenada.

Porgie Rapier had all intentions of going on the excursion with his best friend Ian Hughes. The morning of the excursion, Porgie woke up with a very severe cold. Realising that he could not go on the excursion in this condition, he went to the boatside at about 10 a.m. that morning and sold his ticket to a man he did not know. Porgie, nevertheless, cloaked himself up and went with Colin McIntyre up the west coast of Grenada in *The Invader* in the late afternoon of that day, following the *Island Queen*. Wilfred Jacobs was a good friend of Jerome Alexander, who had all the intentions of making the trip with Jerome, and had already packed. No sooner than he did this, Wilfred's sister unpacked the bag and said he was not to go. She said that he should put the money he was going to spend on this trip towards the rebuilding of the family home that had recently burned down. Feeling guilty, Wilfred stayed at home. [165]

Chykra had looked forward to sharing the trip to St. Vincent with a few of his friends. He invited Colin McIntyre, one of Grenada's most knowledgeable and experienced seamen, who often sailed with him for pleasure. For a reason now forgotten, Colin declined to go, although on the day of the fateful sailing, he followed the *Island Queen* in his boat as far as Beauséjour. H. H. Pilgrim, known as "Box", the father of Arthur

---

165 Later in life Wilfred Jacobs was knighted and also became the Governor of Antigua.

and Norma Pilgrim and godfather to Chykra's niece, Joan Parris, was also invited. But "Box" had promised to play bridge with Sam Brathwaite, and regretfully had to decline. Robin Renwick was another friend who could not go. David Otway had to decline the invitation because he had promised to do some work on his father's estate. All these and others of Chykra's friends would normally have been with him on his very last voyage. It was also unusual that André, Chykra's brother, his brother-in-law, Osborne Steele, and several other of Chykra's usual crewmates were not on this trip. Others were taken in their stead.

Some due to travel on the *Island Queen* changed their tickets to the *Providence Mark* beforehand or on the day of departure. The All Blacks Football Club had been the first to agree to transfer as soon as it was known that the excursion was too large for just one boat. Rosemary "Sylvie" Charles, Beryl Charles and Phyllis Osborne were three of these who originally held tickets for the *Island Queen*, but transferred to the *Providence Mark*. They were all going up to St. Vincent to stay with Bertie Charles, Rosemary and Beryl's brother, who was a doctor stationed there. When the younger people approached them to swap, they readily obliged, as it really did not matter to them which boat they travelled on. Phyllis Osborne forgot to transfer her suitcase to the *Providence Mark*, and when the boats began to pull out she became concerned at this omission. Her companions told her not to worry because it was sure to be taken care of and she would get it in St. Vincent. Roslyn Brathwaite [166] was also going to St. Vincent to visit Bertie Charles, who was her cousin. She had bought a ticket for the *Island Queen*, but her brother suggested that she change it as the *Island Queen* had been having engine trouble. The trip was a gift for her seventeenth birthday and Roslyn was very happy to travel in company with her cousins on the *Providence Mark*.

---

166   Mrs. Solomon Azar

A memorial plaque in the Blessed Sacrament Church, Grand Anse that commemorates both the victims of the *Island Queen*, and those of the *City of St. George*. This is the only memorial anywhere at the time of writing commemorating the lives of the victims of these marine disasters.

PORTRAITS OF CAPTAIN CHYKRA SALHAB AND 17 OTHERS OF THE 67 PERSONS WHO LOST THEIR LIVES ON THE *ISLAND QUEEN*

Septimus Alleyne
Photo courtesy Margaret Dowe

Lucy De Riggs
Photo courtesy Joan Sylvester

Rita Gomas
Photo courtesy Cecil Batholomew

Sylvia Gomas
Photo courtesy Cecil Bartholomew

Humphrey Parris
Photo courtesy Joan Williams

Claire Paterson
Photo courtesy Sir John
and Lady Dorothy Watts

Lucy Paterson
Photo courtesy Cosmo St. Bernard

Redvers Rowley
Photo courtesy Robert Rowley

Doreen Scoon

Laynika Scoon
Photo courtesy Winifred Hercules

Dawn Slinger
Photo courtesy Valerie Steele

Denise Slinger
Photo courtesy Valerie Steele

Joslyn St. Bernard
Photo courtesy Cosmo St. Bernard

Reginald St. Bernard
Photo courtesy Cosmo St. Bernard

Ernest Williamson
Photo courtesy Alice McIntyre

The daughters of the Hon. Alex Fraser and Mrs. Erene Fraser of St. Vincent.
Patsy and Jean, the two youngest, were lost on the *Island Queen*.
Photo courtesy Louie Blencove

Captain Chykra Salhab
Photo courtesy Elinor Lashley

SOME OF THOSE WHOSE LIVES WERE SPARED BECAUSE FOR ONE REASON OR ANOTHER, THEY DID NOT GO ON THE EXCURSION

Rosie De Souza

Alma Murray Kent

Thelma Knight Phillip

Esmai Lumsden Scoon
Photo courtesy Sir Paul Scoon

David Otway   George "Porgie" Rapier

Norma Pilgrim Sinclair

SOME PASSENGERS OF THE *PROVIDENCE MARK* IN LATER LIFE

George Williamson

Selby Donovan
Photo courtesy Nellie Payne

Rosemary "Sylvie" Charles

Beryl Charles
Photo courtesy Enid Charles

Edward Fleming Dowe
Photo courtesy Reginald Dowe

Phyllis Osborne

Lincoln "Jack" Baptiste

Roslyn Brathwaite Azar

Cosmo St. Bernard
Photo courtesy Modern Photo Studio

V.E. Day Parade
9th May, 1945

British Submarines visit St. George's Harbour after the War.

Edward Dowe was Marjorie Baptiste's uncle. He was booked to travel on the *Providence Mark*, and on the day of the excursion sat patiently, waiting for the boats to depart. He saw his niece on the *Island Queen* in all the confusion and put enough pressure on her to change boats and travel with him on the smaller vessel. Margaret Phillip, his future daughter-in-law, had three uncles who were regular crewmen on the *Island Queen*. On the *Island Queen*'s very last trip to Trinidad, two of the uncles, Cuthbert and Allan, had remained in that island to seek other work. Septimus Alleyne, the third uncle, perished.

Selby Donovan held on to his place on the *Island Queen* until his friends of the All Black's Club beckoned for him to join them on the *Providence Mark*. He was slated to sing at Kypsie's wedding, but on the morning of the excursion, he was exhausted. The invitation to join his friends on the *Providence Mark* sounded particularly good, because the crowd on the *Island Queen* was already partying and would party all the way up. Despite always loving a good time, on this particular occasion, Selby decided that it would be better to sleep on the other boat than party, because he had to perform at the wedding and wanted to ensure that he would be able to do this well. This decision saved his life.

There were several others who intended to be on the excursion but did not eventually make the journey. Alma Murray, another of Grenada's really beautiful young ladies, worked at Cable and Wireless. She was going to Kypsie Cruickshank's wedding but, the day before the excursion, she had to cancel her passage when she got the news that her superior officer had been hospitalised, and she had to be in the office in his absence. She remembers her bag was already packed and by the door, ready to be picked up in a hurry to meet the boat. Another person missing the excursion because of someone else's illness was Rosie DeSouza. She was one of the first to book a passage on the *Island Queen*, using the opportunity of the excursion for a vacation in St. Vincent. When the time came, however, her eldest sister was sick in hospital and begged her not to go and leave their ageing mother alone because "these are the days of the war."

Willan E. Julien had considered going on this excursion with a number of other men who were patrons of *Hotel Antilles* but, for some reason, none of this group went. Arnold Fletcher [167] changed his mind about

---

167 Husband of Mavis Fletcher

the trip at the last minute. Esmai Lumsden [168] had been invited to spend the holidays with Merle Minors, one of her schoolmates from the St. Joseph's Convent, and planned to travel on the *Island Queen*. However, Esmai's mother explained that there wasn't time to get permission from her father, who was away. She would be allowed to go on another occasion, provided that he gave his approval. Norma Pilgrim survived because her mother persuaded her to wait a year to go with her friend Denise Slinger to St. Vincent for the holidays. Derek Renwick survived and Ian Hughes did not because they had exchanged tickets, Ian using Derek's ticket for the *Island Queen* so he could travel with his girlfriend, Honey Rapier, while Derek travelled on the *Providence Mark*. Maxim "Bing Bing" Mauricette missed the excursion for the best reason in the world – his wife went into labour and he took her to hospital to await the birth of their daughter Jennlyn.

John Watts was hoping to come up from Trinidad to go on the excursion to St. Vincent. However, he was not granted leave to make this very special outing. S.A. Francis, the owner of the nearby *Hotel Antilles*, had repaired to his hotel when the *Island Queen*'s departure was delayed, and asked to be called when the boat was ready to leave. He whiled away the time playing cards with friends. The summons to board arrived in the middle of a game and, wanting to finish, he sent a message to Chykra asking him to hold the departure until he could get there. The story goes that Chykra told one of his crew to take Mr. Francis's luggage off the boat and leave it on the wharf, because the boat's departure was already so delayed that he would not be waiting for anybody. S.A. Francis, on hearing that the boat was leaving without him, is reported as having said: "Man, let it go! There's plenty of time to get to St. Vincent". [169] Hyacinth DeRiggs got the place that had been kept for S.A. Francis. Anthony DeRiggs recalls that there was the sound of great jubilation as Hyacinth joined his friends.

There had been many others on the pier rejoicing with those lucky enough to get a passage on the two boats, or sad because their wish to go had been thwarted for one reason or another, including being strapped for cash. These later mourned those they had thought lucky, while thankful that they had been spared. Many of those who "almost

---

168 Esmai was first Mrs. McNeilly. Some years after her husband's early death, she married Sir Paul Scoon.

169 Redhead, P. 17

went" suffered for the rest of their lives from "survivor's guilt", a phenomenon common to those who are spared in tragic circumstances while others are not.

Painting by Susan Mains

XVI: SO WHAT REALLY HAPPENED TO THE ISLAND QUEEN?

From the day that the *Island Queen* went missing, there was talk and speculation as to what had happened. Possibilities dwindled as the days and weeks passed, and the boat was still not found. Now there were only a few still open as to the fate of the boat and her passengers.

From the day after the *Island Queen* went missing, a strong rumour began circulating that the *Island Queen* had been sunk by mistake by a British submarine, warship or launch [170], and that all on board were dead. Esmai Lumsden and Margaret Phillip both independently remember the loss of the *Island Queen* being discussed in their hearing. Was the *Island Queen*'s tragic end an error of war, resulting in the loss of life of a boatload of civilians? This is still strongly believed. The sound of *Island Queen*'s German Krupps engine could indeed have made the vessel a mistaken target of a British submarine, warship or armed launch. Because of the delay in the departure of the *Island Queen*, she was where she was not supposed to be, in areas where the U-boats liked to surface, in the middle of the night, when the U-boats were known to prowl. Respondents also report with confidence, but without attendant proof, that personnel of the attacking vessel called for advice on discovering their mistake. It is also commonly believed that boats of various kinds, including the *Harriet Whitaker*, a very fast motor launch belonging to the American Consul stationed in Grenada, had been secretly dispatched before the *Island Queen* was even missed, making a hasty but thorough clean-up operation, netting every bit of evidence they could find. Therefore, when the local population went out looking for the *Island Queen* much later, there was nothing left to find.

---

170 See Grenada National Museum, *Grenada in World War II 1939 – 1945,* which states that the *Island Queen* was sunk by a British submarine. See also Kay, P. 143.

Paula Julien and her mother travelled to Trinidad on the very next trip the *Providence Mark* made after returning from the fateful excursion to St. Vincent. Paula remembers that the talk on the *Providence Mark* among the crew was that the *Island Queen* had been torpedoed by accident, and that all the debris had been picked up. Thelma Knight recounts that the talk at the time was that the British knew that there was a German submarine still in Caribbean waters. A British naval vessel in the Grenadines heard the German engine of the *Island Queen* approaching around midnight and blew up the *Island Queen*, thinking that he was taking out the last U-boat. When the mistake was discovered, every effort was made to cover up the disaster because the British did not want the people in Grenada to know who was responsible for destroying so many of the sons and daughters of the islands, including the children of prominent and influential people.

The case for accidental destruction by friendly fire is strengthened by documented cases where a schooner was mistaken for a submarine, and a submarine mistaken for a schooner. The conning tower of U156 had been mistaken for a schooner when Achilles first entered the harbour of St. Lucia, and on 5th August, 1943, Mariner aircraft out of Trinidad had dropped depth charges in error on a schooner that they had mistaken for U615, which they had been hunting in that area of sea. Kelshall relates that

> To the horror of the crew instead of a broken U-boat, they saw a two-masted inter-island schooner rocking in the waves created by the explosions. Fortunately the schooner did not appear to be sinking but the crew must have been absolutely terrified.[171]

Despite the statements that the *Island Queen* disappeared without a trace, others contend that this is not true and that bits of the *Island Queen* were collected from the sea, or washed up on the beaches. Another rumour was that the *Island Queen* had been shelled and sunk, and that parts of the boat had been brought back to Grenada by the authorities "all shot up". The parts were said to have been concealed in the Government warehouse on the pier and kept there for years. The "dog house" on a schooner is a small, low structure on deck where

---

171 Kelshall, P. 383.

a passenger can take refuge from the sun or where cooking could occasionally be done. The *Island Queen* had two "dog houses", one of which was used exclusively for cooking. Thelma Knight was told by stevedores who frequented her aunt's shop on the Carenage that one of the "dog houses" belonging to the *Island Queen* was brought back intact and locked away on the pier. The stevedores and men working on the pier had witnessed it brought in while the searches by the Fleet Air Arm and the *Rose Marie* were still in progress. The story is that all the remainders of the *Island Queen* including the "dog house" remained in the warehouse on the pier until the pier and all its contents were swallowed by the raging seas during *Hurricane Janet*, which devastated Grenada in 1955.

Ruby DeDier was another who remained sure that some of the items from the *Island Queen* escaped the alleged cleanup operation. She said that people on the west side of the island found items such as hats and other finery, which they believed came from the *Island Queen*. When these were found, they were turned in to the authorities and never heard of nor seen again. Others claim that among the items turned in was one of a pair of watchekongs found on a beach near Grenville. A friend of Donald Linck wondered if it was his. It was also claimed that Chykra's shoe had been found. Fishermen saw debris floating in the sea two or three days later, which they thought had come from the *Island Queen*. The claim that the wood found belonged to the *Island Queen* is based on the fact that the wood was painted black. There are claims that debris belonging to the *Island Queen* was found as far away as Trinidad and Venezuela. One Carriacouan is supposed to still have a piece of the *Island Queen* in her house. But no one claims to have found the part of the *Island Queen* with the name or part of the name, or anything marked with the name of any of the crew or passengers.

The second most popular theory circulating at the time was that a German submarine sank the *Island Queen* without any warning to the passengers. This possibility had already been hinted at in *The West Indian* newspaper of 10th August, 1944, and could have been seized on by readers and those influenced by them at a time when explanations for the tragedy were being frantically sought. It probably brought a modicum of comfort to blame the tragedy on the enemy. However, at this time, most of the German U-boats had either already been sunk or had been called back to base and reassigned out of the Caribbean.

In order to pin down the anti-submarine defences, Germany kept a minimal presence of U-boats in this area. Kurt Lange and his U530 was the last U-boat in the Caribbean area. Between July and August 1944 he remained in the Trinidad area

> "Cruising down the bauxite route, then back up to Trinidad and westwards to Curaçao, but no one knew he was there. There were no radio transmissions, and the coastal radar stations reported the passages clear. [172]

No one knew where he was, and U530 left the Caribbean late in August 1944. Could he have been in the Grenadines charging his batteries and exchanging his air when the *Island Queen* was detected, and he decided to take one last British ship down? It is possible, but the *Island Queen* had a German engine, so how was Lange so sure this schooner was British in the midnight darkness? Would Lange have taken the risk of exposing his position and alerting naval patrol boats in the area by using his guns or torpedoes? Finally, shelling the boat without giving the passengers a chance to alight in lifeboats was atypical of U-boat captains, and there was almost always evidence of a boat meeting its fate at the hands of a U-boat. Compare the instance where 16 bodies were washed up on the beaches of the Carib Reserve in Dominica. These were sailors from a Spanish ship that was blown up by mistake by a U-boat 60 miles east of Dominica earlier in the war. The German captain, in apologising for the sinking of a neutral ship, claimed that he had shot this ship in error, believing it to be a British merchant ship in disguise. [173] But in this instance, there were no bodies to testify to the end to the *Island Queen*. The *Island Queen* may have been sunk by Lange, but this sinking was not claimed by him. Because of the circumstances, it is possible that he did so, but not probable.

The third theory was that there had been an explosion on board the *Island Queen*, resulting in a deadly fire. Carriacouans claim that some wreckage from the *Island Queen* was picked up by the islanders, and these showed evidence that the boat had burnt. Wilfred A. Redhead was the Acting District Officer stationed in Carriacou when the *Island Queen* went missing. He asked people what they thought had happened to the *Island Queen*.

---

172 Kelshall, P. 211
173 Honychurch, P. 171.

> They all came up with the same answer: "She was burnt". When I asked "How do you know?" The answer I got was "Ask the people of Union Island." It appears they saw something burning out to sea, sometime around midnight, from the extreme western point of the island, on Saturday night. When asked why they didn't tell the authorities, they replied, "Well, nobody asked us". [174]

The hospital in Carriacou is situated on a high hill over the town of Hillsborough. The front of the hospital looks out over Hillsborough Bay, while from the back of the hospital several of the Grenadine Islands can be seen, including Union Island and Canouan. There were reports by the nurses on duty that night that there was a loud explosion almost at midnight, and when they looked north they saw smoke and a glow in the direction of Union Island. Some reports say that the nurses also heard faint cries for help coming from that direction. This may have been possible despite the distance, if the cries were borne on the wind. There are people who remember that this explosion was also heard in Sauteurs. However, a fire on board or an explosion would also assuredly have left survivors to tell the tale. The *Island Queen* was equipped with lifeboats, and Ian Hughes and Doreen Scoon were two very strong swimmers on board who could have swum to shore, or survived in the water for a long time.

A very plausible explanation for the loss of the *Island Queen* was that the *Island Queen* had been struck by a floating mine, and exploded with everybody and everything on the boat being torn apart, and all the people killed instantly. Opinions on this differ. Ivan David explains that whereas he does not know what really happened to the *Island Queen*, it is unlikely that a wooden boat casually bouncing into a floating mine would cause it to explode. In Carriacou, a French *Briquette* mine received quite a bit of punishment without exploding, being played on by children, and rolled from one house to another and back again across the sand. [175] It would seem that it takes more than a casual bounce by a wooden boat to explode a floating mine. Grenadian mariners say that this is one of the reasons why those engaged in inter-island trade returned to using wooden schooners during the war. However, Gaylord Kelshall, an authority on the war in the Caribbean, is a proponent of the

---

174 Redhead, P. 19
175 See P. 209

theory that an encounter with a floating mine was the reason for the loss of the *Island Queen*. [176] Kelshall's theory probably influenced Elizabeth Nunez to write in her novel:

> The Americans wasted no time. They laid down mines. La Remous fought back. Cables snapped like dry twigs in her powerful hands. Mines broke free and curled down her whirlpool. Days later La Remous vomited them in the Atlantic. Some drifted as far as Cuba. Ships and schooners exploded, not all of them the enemy's. One, a wedding party from Grenada. Blown to bits, not a single person spared".[177]

It was inescapable fact, not fiction, that floating mines were now a very real hazard to boats in the Caribbean. [178] There is no proof that wooden schooners could not be their victims.

Reports of an explosion, fire and smoke would fit the explanations of a torpedo attack, deadly contact with a floating mine or a fire on board the *Island Queen*. Whereas a fire would surely have left evidence, a torpedo or mine of the size designed to destroy a submarine or large vessel would have also killed all on board the *Island Queen* instantly, transforming all and everything into small pieces that either sunk, were quickly dispersed by wave action, or otherwise disappeared long before the first search party arrived.

A theory that was cited at the time as a possible explanation for the loss of the *Island Queen*, but which on examination does not seem very plausible, is that the *Island Queen* had been a victim of the "Kick 'em Jenny" undersea volcano situated between Grenada and Carriacou. Many people, including Bertram "Molly" Arthur, Julian Rapier and Alister Hughes [179] report that from the very beginning, people wondered if Kick 'em Jenny had been responsible for the disappearance of the *Island Queen*. There was folklore that this undersea volcano would erupt every ten years, and it was time. Dr. Jan Lindsay of the Seismic Unit of the University of the West Indies [180] recently also postulated that

---

176   Kelshall, P. 444.
177   Nunez, P. 170.
178   See section on floating mines.
179   Hughes, P. 62
180   Lindsay, Jan, John Shepherd and Lloyd Lunch. *"Kick 'em Jenny Submarine Volcano: A discussion of hazards and the new Alert level system"* Paper given at the Grenada Country Conference, University of the West Indies, University Centre, Grenada, 7th–9th January, 2002.

if the *Island Queen* had passed over Kick 'em Jenny while the volcano was producing methane gas, the resulting change in the water density would have sucked the *Island Queen* with all its passengers down into a watery grave at the foot of the volcano, with no hope of escape, and no debris to indicate what had happened. However, the *Island Queen* used the "steamer route" far away from the volcano. The *Providence Mark* travelled nearer to Kick 'em Jenny, but it is unthinkable that responsible captains Chykra Salhab and Mark Hall would have deliberately taken this risk of passing near the site of the volcano with schooners full of passengers. Even when Kick 'em Jenny is quiescent, the area of the volcano was known to be a danger to boats with its extremely turbulent seas and strong currents.

A favourite theory of yachtsmen is that a waterspout, rogue wave, whale or freak cyclone had upset the boat. The weather was reported as generally fine, but both Alister Hughes [181] and Cosmo St. Bernard [182] mention that a squall struck the *Providence Mark* shortly after midnight, bringing with it reduced visibility from the blinding rain and wind, preventing the passengers on the *Providence Mark* seeing the lights of the *Island Queen*. At the time, no one was worried about the squall, as such squalls were quite usual. However, a patch of localised bad weather known as a "micro-burst" could have caused the destruction of the *Island Queen*. A "micro-burst" is defined as a sudden violent weather system, almost impossible to detect, exhibiting very strong winds, sometimes in excess of 100 knots, over a small localised area, rarely more than two miles across. This phenomenon has always been known to sailors, but was only recently documented, studied, and given a name when a number of plane crashes and disasters at sea were attributed to this phenomenon. Three of the boats lost at sea were the *Pride of Baltimore*, a clipper, lost in the northern Caribbean near the Bahamas in May 1986, the sail-training ship, the *Marques*, lost a few years before off Bermuda, and the loss of an excursion boat, the *Scitanic*, on the Tennessee River. But in all these cases there were survivors who lived to tell their experiences. [183]

A few people say that the *Island Queen* could have self-destructed. The *Island Queen* had constant problems with its engine and it was

---

181   Hughes, P.45
182   St. Bernard. See Appendix III
183   John Campbell. Campbell, John. "Pride Before the Fall" in Yachting World, March 2010.

rumoured that the propeller sleeve was cracked. The theory was that the engine of the *Island Queen* broke down and the boat drifted, taking in water. Running into bad weather, the boat disintegrated and the debris was scattered by the tides. This theory is mischievous and has little merit because on occasion when the *Island Queen* did break down, it did not take in water, and the boat with its passengers returned safely after landing up at Carúpano, Venezuela. Unless becalmed, the *Island Queen* could travel quite fast under sail alone. Furthermore, the *Island Queen* had just returned from Trinidad, having had a complete overhaul. In any case, the *Island Queen* had lifeboats. This theory was dismissed as of no merit during the Official Enquiry.

There was a further rumour that the *Island Queen*'s ballast had been removed, making it more susceptible to overturn in bad weather. This was also mischievous and unsubstantiated. The Official Commission of Enquiry mentions how much ballast the *Island Queen* would have been carrying. If anyone knew for sure that the ballast had been removed, this certainly would have come out during the Enquiry. This rumour, like the previous one, was circulated to cast aspersions on Chykra Salhab and on the lack of thoroughness by the harbour master, whose duty it was to inspect the boats leaving the island.

All of the above are rational theories as to what happened to the *Island Queen*. In addition to these, there were many fantastic tales that emerged as to the "truth" about the disappearance of the vessel. The most common went like this:

> German spies had discovered that the *"Island Queen"* had a German engine similar to that on Hitler's submarine. They also discovered (and relayed this information to Hitler) that the excursion to St. Vincent was taking place that Saturday.
>
> Sometime during the night, as *"Island Queen"* sailed for St. Vincent, she was captured by the German submarine with Hitler on board. Sailing westwards into the Atlantic, she was repainted, and, by the next morning her rigging had been altered beyond recognition.
>
> Then, with her passengers reduced to the level of slaves to their German captors, *"Island Queen"* sailed south to Argentina. There, Hitler was received secretly by Germans living in that country and then smuggled inland with his slaves to a remote hide-out. [184]

---

184 Hughes P. 62

This story is still believed by some surviving families, who believe that their loved ones are in the Patagonian region of Argentina and, if a search was made, they and their descendants would be found. There are two other versions of this story. One contends that Hitler had trained lady pilots who, at the crucial hour, flew Hitler to Martinique. Meanwhile the *Island Queen* was captured by other agents and repainted like a hospital ship. Hitler joined the disguised *Island Queen* and everyone sailed to Brazil. The other version is that Chykra Salhab was taken by a U-boat because it was thought that he would be able to show the Germans the way into some of the harbours with which the U-Boat captains were unfamiliar. This tale does not say what happened to the passengers.

Still another story was that the *Island Queen* drifted to Africa and that people were seen there who could have been descendants of the survivors. However, the *Island Queen* would have had to drift all the way contrary to the currents.[185] Finally, it was said that "Boysie" Singh, a Trinidadian of infamous fame, was supposed to have captured the *Island Queen*, killed everyone on board and thrown their bodies into the sea. He then repainted and sold the *Island Queen*.

The stories of Hitler, Brazil and Patagonia possibly originated in the reports of what happened to two U-boats that disobeyed the order to return to base at the end of the war. U530, captained now by Otto Wehrmuth, and U977, captained by Schaffer, both well-stocked with fuel and rations and equipped with the latest snorkel device, sailed from Germany to Argentina, where they surrendered. Argentina had a strong pro-German element in the population, and the captains hoped to find a gentler reception here than in defeated Germany. U530 arrived in Mar de la Plata in July 1945 and U977 arrived in August at the same port. They were eventually handed over to the Americans, but not before a hero's welcome was given to the two captains. An inaccurate sensationalist newspaper report which came out in the Argentinean press is summarised by Kelshall as follows:

> U530 and U977 had been part of a ghost convoy which had brought Hitler, Eva Braun and Martin Bormann, plus Nazi treasure to Patagonia and put them ashore before surrendering the boats.

---

185 For a similar belief see footnote 156 on pg 164

Kelshall goes on to say that because the Russians were very quiet about what they had found in Berlin, British and American intelligence took this story very seriously. Consequently, he says "Wehrmuth and Schaffer found themselves very important prisoners indeed". [186]

Any tale was allowable because, as Clayton Steele succinctly puts it, "There is nobody to present themself to say I was there."

XVII: THE OFFICIAL ENQUIRY

When the Grenadian public heard that the Government would hold an official enquiry into the loss of the *Island Queen*, they expected that the most popular theories would form the meat of the official enquiry. The expectation was that the enquiry would finally decide the cause of the loss of the schooner and passengers.

On the 9th October, an enquiry into the loss of the *Island Queen* under the Wireless Ordinance, Cap 249, was agreed to in the Executive Council of Grenada, and one of the other items on the agenda was to find a commissioner. Several members of the Executive Council argued their way out of conducting the enquiry. In the minutes of meeting of 9th October, Mr. E. Gittens Knight said that it was not proper for him to head the enquiry because he had been Acting Colonial Secretary at the time that the vessel went missing. He suggested that the Registrar should be asked instead, and failing him, Henry Steele, Esquire. In the minutes of meeting of 16th October, it was reported that the Registrar later intimated that he was too busy to undertake the enquiry and, in the end, Henry Steele agreed. Magistrate Henry Steele had a personal interest in the proceedings. Thelma Steele, his daughter, had been a passenger on the *Island Queen*. His son, Osborne Steele, had been Chykra's friend, colleague and mechanic, and both Henry and Osborne had lost in-laws and friends on the *Island Queen*'s last voyage.

Cognizance was taken of the fact that the enquiry could not be held under the Wireless Ordinance because this Ordinance only provided for an enquiry relative to a British ship. As Chykra Salhab was not a British

---

186 Kelshall, P. 453.

subject, the *Island Queen* was not registered as a British ship. In the circumstances, it would be necessary to have a commission under the Commission of Enquiry Ordinance.

The Official Enquiry opened on 25th October, 1944. *The Vincentian Newspaper* of 11th November reported that:

> With the cooperation of the Senior British Naval Officer, the Administrator of Grenada has appointed His Worship Henry W. Steele Esquire, a Grenadian Barrister-at-Law and Lieutenant Commander R.G. Liveing, R.N. to be commissioners to enquire into the presumed loss of the vessel named.

Liveing was a specialist in emergencies. Henry William Steele had served with integrity as a Police Magistrate, and was known for his concern for the welfare of the community. He was now seventy years old, and was mild-mannered and well-respected by all classes of Grenadians. His father, John Louis Steele, was a contractor who had lived on Tyrrel Street, St. George's. However, the indications are that the questioning of the witnesses was mainly done by the young, energetic and specially trained Lieutenant Commander Liveing.

The enquiry sat between 25th and 30th October, 1944, and the report prepared at its conclusion was dated 31st October, 1944.[187] The Commission had two mandates: to examine the evidence and come to a conclusion as to the cause of the loss of the *Island Queen*, and to explore and to make recommendations on measures that might be taken to lessen the risk of such a disaster occurring in the future.

To the consternation of the general public, it was discovered that the Commissioners would not examine any of the following explanations as possible or likely causes of the loss of the *Island Queen*: direct enemy action, collision at sea, stranding i.e. being wrecked on rocks, shoals, and internal or external explosion, including the striking of a floating mine, or loss by fire. This list precluded most of the popular theories of what had happened to the *Island Queen*!

---

[187] GOVERNMENT OF GRENADA: Report of the Commissioners appointed to enquire into the supposed loss of the Schooner *"Island Queen"* which cleared the Port of St. George, Grenada, on the 5th day of August, 1944. 31st October 1944 ...Page 1 Section 2. Henceforth cited as The Report

Not surprisingly, the number of people who were willing to come before the Commission and to say what they knew or suspected, was deemed disappointing. Lack of response from the populace was hardly surprising. Grenadians are naturally reticent in matters concerning giving information to the authorities, especially if it was determined that circumstances might not change as a result. Giving evidence would not bring back anyone from the dead. In addition, people knew that in wartime, and because they could not prove anything, expressing themselves might get them in serious trouble, especially as discussion on certain likely causes of the loss of the *Island Queen* had been already excluded from the proceedings, and would certainly have been detrimental to Britain and her Allies.

Grenadians had not forgotten the imprisonment of Owen Wells in Grenville in 1942 for expressing an opinion that was thought by some to be seditious. Many also recalled the incident involving a preacher in Barbados, whose case was reported in *The West Indian* for Tuesday 9th July, 1940. Under a headline, "Barbados Wayside Preacher Imprisoned", the news item, citing the *Barbados Advocate* as the source, stated that:

> Frank Beckles, a "preacher" of Holetown, St. James, was yesterday ordered by Mr. W. K. Ferguson, City Police Magistrate, to undergo one month's imprisonment with hard labour on the charge brought against him by L/Sgt. Simmonds for that he at the parish of Christ Church, within the jurisdiction of Section "A" on June 6, 1940, did orally influence opinion in this island in a manner likely to be prejudicial to the efficient prosecution of the war.

The Commissioners were aware of disgruntled public opinion about the enquiry, for they lamented in their report that:

> In the course of our enquiry, it was noticeable that there was considerable reluctance on the part of persons to come forward and assist us with our investigations. This reluctance did not extend outside the official proceedings where rumours, often of an injurious and fantastic nature, were widely current and widely believed. [188]

There was nevertheless a keen interest in the proceedings of the enquiry. Every day the courtroom was full of spectators, among them T. A. Marryshow. Emotions ran high. Marryshow was seen crying

---

188 *The Report*, Section 3.

at the enquiry, undoubtedly not the only person to be once more overcome with sorrow.

A total of thirteen witnesses testified before the Commission on the two issues that The Report was instructed to examine: Messrs. George Banfield, Henry Gun-Munroe, Mark Hall, Jocelyn Ireland, Gittens Knight, David Lusan, Colin McIntyre, Beresford Prince, Irvin Redhead, André Salhab, Osborne Steele, and Sydney Wells. The Hon. Terrence B. Comissiong, the Colonial Treasurer, whose post combined Collector of Customs and Excise, Commissioner of Income Tax, Manager of the Government Savings Bank, Registrar of Shipping, and Shipping Master, also gave evidence, but this was confined to the section of the enquiry dealing with recommendations for new regulations to govern local shipping and to ensure a greater measure of safety. Comissiong had been in hospital, and the duties of his post had been temporarily taken over by Gittens Knight when the *Island Queen* disappeared.

It would seem that the Lieutenant Commander favoured the theory that the *Island Queen* was lost when it capsized during a sudden squall. According to a report in *The West Indian*,[189] the following exchange took place between Lt. Commander Liveing and Captain Wells:

> Lt. COM LIVEING: ... Would you agree that the weather experienced by the *Providence Mark* can be accepted to be the same as that experienced by the *Island Queen* if she continued on the voyage?
>
> CAPTAIN WELLS: Not necessarily.
>
> LT. COM. LIVEING: Why not?
>
> CAPTAIN WELLS: A local squall is not very large and if ships were five miles apart, they might experience different weather.
>
> LT. COM. LIVEING: Can such a squall be of high intensity in the circumstances you describe?
>
> CAPTAIN WELLS: Very high.
>
> LT. COM. LIVEING: If such a squall overtook a master unaware, and he had been unable to shorten the sail in time, what would you expect to happen?
>
> CAPTAIN WELLS: The ship would very probably capsize.

---

189 *"Island Queen* Inquiry Closed" *The West Indian.* 31st October, 1944

> LT. COM. LIVEING: If a Schooner of 65 tons carrying 15 tons of ballast with an engine assembly of 7 1/2 tons were to capsize, do you think she would sink at once?
>
> CAPTAIN WELLS: She would sink as soon as all the air from her hull escaped.
>
> LT. COM. LIVEING: In the normal conditions of a voyage in this area with expected fair weather, would hatches be open?
>
> CAPTAIN WELLS: The hatches would be open in fair weather.

And a bit later, this question was asked:

> LT. COM. LIVEING: In the circumstances of a ship of the size of the *"Island Queen"* being struck by an unexpected squall, would the presence of 56 passengers increase the danger in an emergency?
>
> CAPTAIN WELLS: If they were on deck they certainly would.

After examining all of these other possibilities, the Commissioners concluded that the *Island Queen* had probably been lost by "Foundering".

> A ship may founder by reason of capsizing, springing a leak, or being overwhelmed by heavy seas, or a combination of these ... (W)e can eliminate foundering through springing a leak or by reason of heavy seas. There is an abundance of evidence to show that the hull of the *"Island Queen"* was sound and that a defect which had, on a former occasion, led to a leak through the stern gland had since been made good. Also there is satisfactory evidence that the weather in the area at the material time was generally good... We have the support of expert opinion in stating that if the vessel had been struck on the beam by a sudden squall before the Master had time to shorten the sail, then the ship would very probably capsize. The danger would be aggravated if at the time the vessel was almost stopped and a substantial number of the passengers were upon the upper deck.[190]

The following scenario is suggested by The Report: The engine of the *Island Queen* probably broke down around 8 p.m., possibly because the captain was racing the boat to St. Vincent against the *Providence Mark*.

---

190 *The Report*, Page 4 Section 13 (f)

Although the *Island Queen* was an auxiliary schooner, she depended more on her engines than on sails. Captain Salhab was working on the engine below deck and was oblivious of the approaching squall. The squall hit the boat, the boat capsized and water rushed into the hatches that would normally be open. The resulting inrush of water sank the *Island Queen* almost immediately because of the ship's ballast and the massive engine assembly.

The exact statement by the commissioners was: "We conclude that the Schooner '*Island Queen*' must be presumed lost by marine hazard." [191] This conclusion said exactly ... nothing!

The acute dissatisfaction over the handling of the first mandate obliterated from people's mind anything else the Report had to say. As a result, the useful recommendations that came from a discussion of the second mandate of the enquiry are not even remembered as forming a part of the Report. The measures suggested in these parts of The Report did contribute significantly to the safety regulations subsequently implemented governing inter-island shipping and passenger safety. Unfortunately, another marine tragedy had to take place[192] before adequate safety precautions were implemented regulating schooner travel.

There was widespread anger, and accusations were made that the Commission's Report was a cover-up. The relatives of those who were lost on the *Island Queen* were very dissatisfied and bitter, and some were full of suspicion. Could the Commission of Enquiry have been used to suppress the truth so as to prevent a major scandal erupting in the middle of the war? Were the Commissioners unable to ask the right questions? The Commissioners had not examined the most popular theories of the catastrophe, but instead speculated and supposed in their conclusion that the "floundering through capsizing" of the boat was "not only a possibility but a probability". [193] Moreover, The Report insinuated that the *Island Queen* was an inferior vessel and cast aspersions on the captainship of Chykra Salhab, surmising that he

---

191 *The Report*, Sections 21 and 22

192 On 19th June, 1971, the auxiliary schooner *City of George* was returning to Grenada from Trinidad when it caught fire. On board there was no radio, only one lifeboat, and no life jackets or fire fighting equipment. As a result, there was considerable loss of life. Twenty-two people drowned, thirteen of whom were from Windward, in Carriacou.

193 *The Report*, Section 13 f

might have been inattentive to the weather conditions, when both the seaworthiness of the *Island Queen* and the unquestioned skill and sense of responsibility of the captain were well-known.

As long as the *Island Queen* and its passengers are remembered, the truth will always be sought because, for some, the conclusions of the enquiry were unacceptable. Maybe the truth will be found in the official World War II records that are slowly being released for public scrutiny sixty years after the end of the war. The likelihood is, however, that the truth about *Island Queen*'s disappearance and the last moments in the lives of the passengers will never be known. The sea may always keep this secret.

XVIII: NO END TO THE ISLAND QUEEN STORY?

If it was hoped that the enquiry would provide closure on the fate of the *Island Queen* and its passengers, it did not happen. The enquiry provoked more questions than it answered. Because of this, the chronicle of the *Island Queen* refuses to fade away. Over sixty years after the event, there are still people in Grenada who feel a great injustice was done by way of the enquiry.

One conservative opinion of the enquiry was that "It was held just to assemble what was known and to provide a record." The latter part of the statement is certainly true. The Report of the enquiry has become the solitary official reference to those seeking information on the *Island Queen*, and there is nothing more to be learned at this time from the official records. This incident among the very small islands of the West Indies is now all but forgotten by those living outside the Caribbean who are not members of the Caribbean diaspora. On a sojourn in England in the 1960s and 1970s Robby Rowley, an engineer with the U.S. Army, who had lost his father on the *Island Queen*, made enquiries about the incident at the British Admiralty. He could find no one who had ever heard of the *Island Queen*, much less what happened to it. Searches in the National Archives at Kew in London only reveal the enquiry and an entirely different vessel another part of the world that also carried the name *Island Queen*.

Searchers with a plethora of information now available to them on the internet find one of my articles and newspaper excerpts based on it. However, hidden in other websites scrutinising the events of World War II are discussions on what happened to the *Island Queen*. One "thread" even gives the name of the British warship supposed to have been responsible for its destruction.

There are also several persistent and similar stories of sailors or personnel from the armed forces of both the Allied nations and Germany who visit Grenada or St. Vincent for the express purpose of letting Grenadians and Vincentians know the truth about the fate of the *Island Queen*. Jack Baptiste relates that sometime between 1948 and 1950, a German couple came into his bookstore *Seachange*, and asked him if he knew anyone who had travelled on the *Providence Mark* on the fateful August 1944 excursion. He admitted that he did, and was asked if he could find one other person to meet them for a drink at the *Holiday Inn* in Grand Anse. Jack invited Phyllis Osborne to go with him to meet the German tourists. After some pleasantries, the German said that he had been a member of the crew of a U-boat during the war, had briefly seen Grenada and how beautiful she was. He had decided to come back on vacation whenever he could. He told Jack and Phyllis that he wanted Grenadians to know that the Germans had not torpedoed the *Island Queen*. Apart from all other considerations, at this time in the war, Germans would not have wasted a torpedo on a small wooden boat.

Another tale was told by Dr. Earle Kirby of St. Vincent, who was a well-known and respected vet in the Government service. Several years after the war ended, he was summoned to come outside his office in the Ministry of Agriculture building to meet a man who said he had a tale to tell him. The man refused an invitation to come inside, but told Kirby that he used to work at the U.S. Base at Vieux Fort in St. Lucia. The British and US sailors would meet at the base, and swap tales of the war. He overheard a conversation between sailors concerning the *Island Queen*, which indicated that a U.K. boat had sunk the *Island Queen* west of Canouan. When it was realised that a terrible error had been made, the boat collected all the flotsam and took it back to St. Lucia, where it was burnt. But what of the bodies? What happened to them? Were these also considered as flotsam? Were they also burned?

This tale is echoed in the gruesome and disturbing account related by Maurice Paterson of Grenada. In 1945, he says, a man called Clifton

Dottin, who should have travelled on *Island Queen*, but was one of the "lucky escapees", shared a restaurant table in St. George's with a British naval officer. The two talked first about the war, and then the subject of the *Island Queen* came up. In 1992, Dottin, now an old man, told Paterson that:

> The naval officer said to me that his submarine picked up the beat of the German engine on the *Island Queen* and torpedoed it. When they surfaced, the captain panicked and ordered all survivors killed, all debris collected." [194]

In May of 2003, I received a long distance telephone call from Austin Hughes, the brother of Ian Hughes, one of those lost on the *Island Queen*. Austin, who had migrated to the United States, said he had been living with a secret all these years. He was then aged 73, and did not want to die with the secret. He wanted me to let the public know that when he was a youngster of 16 and 17 in 1947 and 1948, he worked at Government House as a Code and Cipher clerk. He said that he had searched for and found the "Top Secret" file on the *Island Queen*. In the file was a telegram apologising to the Governor for the accidental shelling and sinking of the *Island Queen* by a British boat. The *Island Queen*, with its German Krupps engine, had been mistaken for a U-tanker that was thought to be on its way to re-fuel and provision U-boats among the Grenadines. Several people have tried to locate this file on the *Island Queen*, but it has never been found.

Alister McIntyre worked in the same capacity as Austin Hughes between 1952 and 1954. While there, he came across a document that made reference to the loss of the *Island Queen*. This document said that correspondence regarding the loss of the *Island Queen* had been on a strictly personal level with the Governor who was in London at the time of the event, because the British Government would wish to avoid embarrassment if the full details in the personal correspondence were to be revealed. Alister never saw this person-to-person correspondence, but feels that Austin Hughes did. He knew Austin, who he remembers was called "Rifle", and said that Austin was "a very reliable fellow". In spite of these eyewitness accounts of documentation concerning the fate of the *Island Queen*, the "Top Secret" file remains just that: top secret.

Finally, there is recent speculation that the *Island Queen* with all its passengers might have been taken by a spacecraft. The general belief

---

194   Paterson, P. 51

among "UFO buffs" [195] is that aliens seek young genetic material to strengthen their gene pool, which is deteriorating as an unrecognised result of reproduction by cloning. The *Island Queen*, travelling in a lonely stretch of sea in the middle of the night, with all its young, healthy passengers, was ideal for the purpose of obtaining new genetic material. The schooner was taken up, complete with its passengers and crew. The vessel and passengers would have been returned to the same spot after a day or two with the memories of their abduction removed. However, the alarm was given too soon from St. Vincent and the area was promptly swarming with searches by air and by sea. The passengers and crew were therefore retained by the aliens to live the rest of their lives as guests either on an alien planet, or on a "mother ship", in the future, or in a parallel universe. However, this explanation remains a theory, with absolutely no evidence pro or con. The glow that the nurses saw, and the absence of any wreckage or bodies, was explained by the brilliant light that would be emitted by a UFO hovering and a flash as it departed abducting the whole vessel.

Before dismissing this last theory as preposterous, it must be recalled that in the 1970s, the United States of America was about to set up a research station in the Caribbean to study UFOs. The fact that Grenadian Prime Minister Gairy gave his "UFO" speech to the United Nations in 1978 [196] was not as ridiculous as Grenadians were led to believe by the antagonists of Mr. Gairy. It was a manoeuvre to have the UFO research station located in Grenada, with the ensuing economic benefits to Grenada. There are stable, rational individuals in Grenada and in Trinidad who claim to have witnessed UFOs rising out of the sea.

Even if the truth is ever discovered, as long as they are loved and remembered, the unfortunate passengers of the *Island Queen* will live on, and the story will never be told for the last time.

---

195 UFO is the acronym for Unidentified Flying Objects

196 Address of the Rt.Hon. Sir Eric Gairy, Prime Minister of Grenada, to the General Assembly of the United Nations at its Thirty-Third Session on 12th October, 1978.

He too had been eighteen, like most of these.
It was then that the war took on its meaning for me,
a meaning that would never change.
It meant only that people without choice in the matter
were broken and spilled, and nothing could ever take the
place of them.
But I did not think of them continually.
Even at this relatively short distance,
I already only thought of them only from time to time.
It was this that seemed a betrayal.

(From Margaret Laurence *A Bird in the House.*
A Seal Book published by arrangement with McClennand and Stewart Ltd. 1970,
Toronto. Pgs. 169 – 170)

PART VI:

THE WAR ENDS

I: CELEBRATING "V.E." DAY

The American entry into the war marked the turning point for the Allies. Even though there were some moments of concern after December 1941, the tide was turned, and it soon became clear that Germany and her Allies would be eventually defeated. The inevitable happened on 8th May, 1945, when all fighting ceased in Europe. This day, dubbed "V.E. Day" (Victory in Europe) was greeted with relief, joy and celebration.

In Grenada, the voice of Winston Churchill came through on the 7 a.m. BBC news, declaring victory for the Allies or "United Nations" as he had taken to calling them. Loud cries and screams and sounds of rejoicing were heard from radio listeners in many middle-class homes in Grenada, otherwise known for quiet decorum. This drew the attention of others who, upon hearing the cause of the commotion, joined in the jubilation. The bells of all the churches started to chime, and as news of the end of the war spread though St. George's, the streets were filled with people chanting and clapping, jumping up, and celebrating.

There was also music, dancing, shell blowing, pan beating and a carnival-like atmosphere. Relief broke down people's reserves.

Union Jacks appeared all over the place. Drucilla Slinger remembers that otherwise contained businessmen were seen running through St. George's with flags and shouting for joy. People decorated their houses and did anything and everything that they could think of appropriate to their celebratory mood.

The day after the announcement, there was a parade in the Market Square, with the governor in attendance. The Volunteer Reserve, the cadets, the police, the soldiers from the Southern Caribbean Force and those recruited but not yet dispatched of the Caribbean Regiment all paraded for the joyous public. Alister McIntyre remembers that there were some American soldiers here at the time who also participated in the parade. Brenda Wells, who was about eight years old at the time, was taken to see the parade and lifted up by her uncle above the heads of the adults so she could see. Margaret Phillip was also taken to see this parade by her family. On the way they saw Ben Roberts, a family friend who lived in Grenville Street, and they exchanged exuberant greetings fit for the occasion.

The nuns at St. Joseph's Convent thought that the manner of the celebrations in town was neither seemly nor decorous, and was no place for their girls. The girls were kept in school by the nuns, causing Fr. Alfred Pike O.P. [197] to tease the girls, calling them Germans, because they were not celebrating the Allied Victory.

The celebrations in St. George's lasted all day and all night around the Market Square and on the Carenage. Private parties were held in clubs such as the Aquatic Club, Silver Sands Club, and at the Richmond Hill Tennis Club. There were also parties in private homes. One of the best parties is said to have been the one Jack Baptiste held for his friends. Wherever they celebrated, the young people danced until they could dance no more. Rhona Otway [198] remembers a lighted victory sign on the pier. In a few days when the mood was quieter, talk began about what the end of the war would mean – the soldiers coming home, boats being able to travel without fear, the resumption of the shipping of Grenadian produce and imported goods reappearing on the shelves of the shops.

---

197 Fr. Pike transformed stables into the original Roman Catholic Church at Roxborough. This was one of the many churches that was demolished by *Hurricane Ivan* in 2004 and has been completely rebuilt, and rededicated in January 2011.

198 Mrs. Arthur Pilgrim

More rejoicing took place on 14th August, 1945 — "V.J." Day, which marked the Allied victory over Japan, and the definite end of the Second World War, but this day was not as elaborately celebrated. In order to humble Japan and bring that part of the war to a quick end, a Boeing B-29 Superfortress, nicknamed *The Enola Gray,* dropped the atomic bomb, code-named *Little Boy,* on the city of Hiroshima on 6th August, 1945, wiping out its entire population of 343,000. This was followed on 9th August, 1945, by another bomb, code-named *Fat Man,* dropped on the city of Nagasaki from the B-29 Superfortress *Bockscar,* which had similar terrible results. Although Japan had vowed to fight to the bitter end to prevent further suffering of the Japanese people, the Hirohito, Emperor Showa announced the surrender of Japan in a radio address to the nation on 15th August, 1945.

For children living outside St. George's, the end of the war came much more quietly. Gunny Swapp knew that the war had ended only because he heard people talking about it. Many understood something of the implications of the Allied victory. The boys took delight in two new types of sweets given out to them as part of the celebrations. They called them "Hitler's Balls" and "Hitler's Stones".

II: RESULTS OF WAR

A global war had been a new experience for Grenada, with new anxieties, trials, hardships and pain. Little by little, Grenada tried to settle back to its pre-war ways, but life in Grenada had changed. In general, behaviour was more subdued. People did not walk about or "lime" in the streets at night, and they became more home-centred. They also learned to be thrifty, live within their means, and to be satisfied with what was available. Grenadians also had to make peace with lost opportunities.

A group of young people with outstanding ability that lost their opportunity for fame and fortune because of the war was *The Harmony Kings.* This was a small musical band from Grenada that was so talented that it had quickly become quite celebrated, and was in steady demand for engagements in Grenada and the other islands. *The Harmony Kings* were often engaged to play the accompaniment for already famous calypsonians, who were recording for the Decca Label at studios in Trinidad. *The Harmony Kings,* although recognised as a

Grenadian band, had recruited musicians from Barbados, Guyana and Trinidad. Among the members of this band was Dennis Malins-Smith, who played the trumpet, his brother, Bert, who played the saxophone, and Joseph Griffith, a Barbadian. When war broke out, the band had to disband because travel was too dangerous. Although *The Harmony Kings* remain a musical legend, their further fame and development were thwarted by the war.

Those who had been traumatised by events in the war never forgot. In the early 1980s, Rosamond Barker-Hahlo was a passenger on a Geest line boat. She was invited by the German captain to his table for dinner. A premonition made her decline but, in chatting with him later, she discovered that he had been the very same captain of the U-boat that had torpedoed the *Regents Tiger*, causing so much distress to herself, her husband, and the other passengers in August 1939.

The parents, relatives and friends of the three young Grenadian RAF airmen and the Grenadian firewatcher killed in action continued to carry their loss, as did the parents, relatives and friends of those who had been lost on the *Island Queen*. The loss of the *Island Queen* had taken place during the school holidays, and many children had not been in town for the event, having gone home or having been sent to relatives in the country. The first news had been kept from them, but it was impossible that the disaster could be hidden in the long-term. The event made an indelible impression on the children of the nation and, depending on how close they had been to those who died, they carry the grief to this day.

In spite of all these and other distresses brought upon Grenada by the war, the predominant opinion is that the hardships of World War II influenced Grenadian behaviour for the better. There was a change in people's attitudes towards each other, as they learnt tolerance and began to pull together as a community. They became used to being resourceful and careful with scarce manufactured goods in their possession. Elinor Salhab remembers that there was no complaining during the war, and people were grateful for what they had. Much of the values learnt in wartime would remain in the culture for many years.

The aftermath of war found Grenadians healthier than ever. The food available to them was the right sort of food for good health. They got more exercise due to the lack of transportation. People used to regularly walk from Gouyave to Sauteurs or from St. George's to Birchgrove.

There was one gentleman from St. Andrew's who worked in town, who used to walk from St. George's to Grenville each weekend to see his family! If people did not walk, their social interaction would be severely curtailed — so they walked. These benefits born of necessity would be gradually eroded as people took advantage of an easier post-war lifestyle, the return of public transportation and the gradual disappearance of shortages of imported food items. However, while Grenadians ceased long-distance walking, the Grenadian cuisine that developed during the war, which reintroduced traditional dishes and utilised local produce in many different ways, became the foundation of the unique culinary experience that is one of the boasts of Grenada today.

The Americans had begun dismantling their bases even before the war finally came to an end, and very soon there were only ghost towns left where before there had been bustling bases. Permanent items left behind were the airfields and airstrips. Especially in Trinidad, many of these were abandoned, but in the Eastern Caribbean the larger airfields would be developed into international airports. Pearls airport in Grenada was to remain the only airport in this island for 42 years until 1984 when the Point Salines Airport was opened. [199]

British war ships, submarines and aircraft visited the islands of the Caribbean, including Grenada, to say goodbye to a people who had remained loyal supporters of Britain through this time of trial and tension. The public was invited to inspect the war ships and special invitations went out to the schools to have their pupils tour the vessels. Marcella Lashley vividly remembers visiting a war ship in port, and the sailors showing them around. The music bands off the visiting war ships also entertained Grenadians with greatly appreciated concerts at Fort George. Different types of aircraft flew over Grenada demonstrating their might and skill. Children remember the awesome thunder of

---

199 The airport at Pearls soon became antiquated and inadequate to meet the needs of modern air transport because the runway was short, and the surrounding hills did not permit night landing. Land had been put under covenant at Point Salines for the construction of a modern airport. However, the needs of the 1979 Grenada Revolution were such that it was imperative that there be an airport that could accommodate the types of aircraft that Pearls Airport could not handle. Construction on the airport at Point Salines began immediately. The construction was undertaken by Cubans, and paid for by a heavy tax on the Grenadian people. The end of the revolution saw the runways usable, but the facility unfinished. The facility was completed and formally opened in 1984 as the Point Salines International Airport. On 29th May, 2009 the airport was renamed Maurice Bishop International Airport in honour of Grenada's deceased revolutionary prime minister.

the formations of planes as they flew low overhead. A pair of British submarines visited Grenada and, after remaining in port for several days and allowing many to undertake tours on them, they departed. But before finally taking their leave, they put on a "show" in the harbour, submerging and surfacing several times to the wonderment and delight of onlookers.

Tubal Uriah "Buzz" Butler was released from the detention centre on Caledonia Island on 10th April, 1945, just before the war ended. His release was celebrated both in Trinidad and Grenada. He seemed to have had not one ounce of resentment but, as Michael Anthony says:

> If the authorities had counted on five years of internment turning Butler's thoughts away from politics, they were soon to be disillusioned, for Butler told reporters on his arrival in Port-of-Spain: I am looking for a home and an office in Port-of-Spain. I have been keeping abreast of West Indian politics and consider the time has come when West Indians should be allowed to determine their own affairs."[200]

Butler remained at the forefront of Trinidadian politics until Eric Williams became the dominant figure a decade later. Butler died at the age of 80, having achieved most of his goals. He was suitably honoured by having a major new road in Trinidad named after him. Today it is a matter of national pride for many Grenadians that Tubal Uriah "Buzz" Butler, one of the Fathers of Caribbean Trade Unionism, was born, spent his formative years, and was raised to manhood in Grenada.

Carnival had been cancelled during the war years, and in 1946 King Carnival was re-instated in great style, with a King and Queen of Carnival arriving in St. George's by motor launch. But the new Grenada Carnival was a version based on the Carnivals of Trinidad and Rio. Nevertheless, the traditional masquerades refused to die and remain triumphantly undefeated as the defining figures of Grenadian Mas.

The Competent Authority under Gittens Knight remained in force for some time after the war ended. There was a dollar shortage in sterling areas and as a result, Grenadians continued to need licences for a number of still scheduled items until the mid 1950s. Gradually, however, imported goods not seen since the beginning of the war reappeared on the shelves. This put a dent in the pocket of Merry McIntyre, for as

---

200 Anthony, P. 262

soon as imported toiletries arrived, his customers deserted him for the imported products and he had no more sales for his locally-produced items. However, he continued to produce bay rum until his death in 1952, for this was a unique product for which there was no imported substitute.

Right after the war, many new products came on the market, made possible by the research initiated to produce better weapons and war materiel. Many of these materials and inventions were introduced to Grenada. The greatest impact was that of plastics, the introduction of which had both a good and a definite downside, as plastic was from the beginning of its use in Grenada one of the biggest polluters of Grenada and the Caribbean. Also making its appearance in Grenada was the transistor radio, said to play a major part in urbanisation. More products introduced to Grenada at this time were the ball point pen, detergents and shampoos. There were also new tools and equipment for Grenadians to learn to use, which would make life easier and work more efficient. These included everything from power tools to bulldozers.

Adina Maitland remembers that the veterans and those returning to Grenada from Trinidad and Aruba came back with many new ideas and experiences. They had been exposed to different lifestyles, values and world views, and were energised to effect change in Grenada. If Grenadians only came back to visit, even a short stay was enough to pass on something they had learnt overseas.

The war launched Grenada onto a wider world stage, opening new horizons geographically, technologically, and socially. The war altered the way Grenadians thought about their place in the world. Two of the most remarkable social changes were that in the plantation era, Grenadians had been very colour- and class-conscious. Two remarkable realisations were brought about as a result of the war. Firstly that the shades of skin colour, so very important in Grenada, did not matter in many countries, including the U.S.A. and Britain, where there were only two categories: black or white. Secondly, although remnants of the old divisions remained in Grenada, never again would people in Grenada have insurmountable social distances between them. The war also ushered in a new class system based on achievement criteria. The Americans in Trinidad, in spite of the racial attitudes some would have had, generally treated labour with far more dignity than was the custom among the managers of plantations and other traditional employers in

Grenada. For the first time, Grenadians were exposed to contractual labour, which bound both worker and employer, and to an alternative idea of labour relations to the almost feudal arrangements they had left in Grenada. Grenadians who had worked with the Americans in Trinidad came back with a new sense of self-worth.

Grenadians who had been in Britain and Europe had also been treated with more respect there than they had received at home, affording them a glimpse of equality. They had, for the first time, seen white men engaged in menial labour, and had learned not to judge a man's worth simply by the colour of his skin. Although subjected to prejudice, this was not universal, and Grenadians had been welcomed as members of the armed forces and the general society wherever they served in Europe.

Some Grenadians returned with enough money and/or skills to set themselves up in small business. Thus the private sector was increased and was injected with new life. Grenadians who had also been exposed to the "American Way of Life" and life in Europe sought to raise their living standards on their return to Grenada.

Although some of the migrant workers returned home, many did not. Grenadians in Trinidad became a resource for those left at home, both providing a means for the further migration of nationals, and a source of funds through remittances.

When shipments of cocoa and nutmegs resumed, peasant farmers were able to sell the produce they had stored. The stored produce acted as family savings, and the money obtained was used to carry out much-needed repairs on their houses. For the first time, many of the houses that previously had been thatched, were now covered with galvanised metal sheets. Larger farmers, who had stockpiled nutmegs as an investment, were launched by their profits into the realm of the *nouveaux riche*.

III: THE MINE EXPLOSION IN WINDWARD, CARRIACOU

Even after the war, there were problems with war detritus left behind. Immediately after the end of the war, the US military swept the minefields at the entrances to the Gulf of Paria with the object of clearing the remaining mines. What they discovered was that these mines had almost all broken free. None of the mines laid there in 1942 were found in place, and only 10% of those laid subsequent to 1942 were still anchored. They could now be anywhere, drifting about the Caribbean or out into the Atlantic having joined the Gulf Stream. Some mines were discovered hiding among the mangroves of Trinidad and the other islands, and were detonated, but most were never recovered.

In addition to American mines floating in the Caribbean and scattered on different shores, French Briquette mines had torn loose and floated across the Atlantic into the Caribbean all the way from Senegal. Many of these were found in the marshes and salt flats of British Guiana, and were exploded by British bomb disposal units. Some French mines were spotted on the water by schooners and small craft, and reported to the military authorities, who took care of them. It was one of these Briquette mines that washed up on the beach in Carriacou, and was responsible for a terrible tragedy in the little village of Windward in July 1945.

To the villagers, the object looked different from the usual barrels of "come-ashore", and there was some speculation among them that this could very well be a chest of money from a ship that had recently sunk near Carriacou. For many days, there was a tussle between neighbours for possession of the mysterious object. During the night, one family would roll the mine over the sand to rest in front of their house, and the next night another family would roll it back. During the day, it was used to play "horsie" by children. Patrick Compton remembers that he was one of these children who used to ride on the mine. The handle on it was also a source of amusement for the children because, if this was pushed, it started ticking. Pushing it in the other direction would make the ticking stop.

The story is told that some Barbadian masons working on a house nearby had seen one of these things and warned the people that the object was a mine, but they ignored the warning. [201]

The mine lay on the sand for eight days, near a boat under construction which, when completed, would be called the *Amber Jack*. On the ninth day, one family got wind of a plan that the other family had intentions of moving the object that night under a house in the village, and opening it there. The first family were determined that if the object was to be opened, it was to be opened in plain sight of the community, who should all share in the largesse it was thought to contain. On the 6th July, 1945, between the hours of 2 p.m. and 3 p.m., an attempt was made to open the mine with a cold chisel and hammer. It is reported that on the first strike, a loud enough "whoosh" was heard, which onlookers said was the money "letting off stale air". On the second strike the mine exploded, killing nine people instantly, Another shipwright who was walking away from the scene was killed by a piece of shrapnel passing through his neck. Shrapnel flew as far as Old Wall", and several other people were injured, including Headley Enoe, who had a piece of shrapnel lodged in his leg.

There were lucky escapes as well. One young man, Peter Patrice, who should have been on the spot working on the boat, left very late for lunch with his friend Patrick McLawrence. When he got home, there was no lunch. McLawrence urged him to return to work on the schooner, but he said he was not going without lunch. McLawrence returned to work, and was the one of the two friends who lost his life when the bomb went off shortly after his return to the site. Many others narrowly missed being injured or killed by shrapnel flying through the air. Patrick Compton's father Ville Compton was weeding in his garden when a piece of shrapnel whizzed past his head, went through the house opposite to his, and lodged in the ceiling. Had it been just a little closer to Ville's head...

Michael Huton lived in Petit Martinique [202] as a child. On that fateful afternoon, his aunt had called him "to run a message" for her. As he

---

201 The story goes that when the Barbadians insisted that "It's a mine" the Carriacouans told the Barbadians in no uncertain terms — "No, it's not a-yours, it's a-mine". If this really happened it was a most unfortunate misunderstanding.

202 Petit Martinique is the third island of the tri-island state of Grenada. It faces the village of Windward, Carriacou.

turned to leave the veranda, he saw a column of black smoke rise up across the water from Windward. Then he heard the tremendous blast. He remembers that people rushed to go across to Windward to see what had happened and were amazed at the enormous crater the blast had left.

Cosmo St. Bernard, who was visiting Carriacou on the day of the blast, heard what he aptly calls a "hellish explosion" from his location in Hillsborough, the capital of Carriacou, about two miles away. He was among the many people who hurried to Windward to see what could have possibly happened. Edward Kent also hurried to the scene of a terrible tragedy. What they saw resembled a war zone. When the powerful mine had exploded, everything in the vicinity had been completely flattened and laid waste, including all the large sea-grape trees. The nearest victims to the explosion had had their bodies ripped to shreds, with human remains scattered over the area. Other bodies were strewn about the scene in various stages of dismemberment. Sadly two little girls, Marjorie and Cicley, also died.

> Two of the victims were small flaxen-haired girls who were collecting wood chips from near where a vessel was being built. [203]

In the absence of coffins or stretchers, Kent suggested that doors be removed from nearby houses to substitute. The bodies which were more or less intact, as well as remains identified from the other victims, were gathered up and carried on these to their homes. Three boats came from Canouan *post haste* to take the children from that island attending school in Windward back home, because there were fears that something else would happen, The Briquette mine had done enough damage, however, and thankfully there was nothing else to follow except that the populations of Windward in particular and Carriacou in general were thoroughly traumatised.

As bad as the disaster was, it could have been much worse. The crater the bomb left was measured at a diameter of eleven feet, but most of the blast went out to sea, and the sand absorbed some more of the shock. What if the mine had been taken into the village?

A plaque has recently been erected on the spot of this tragedy by the Ministry of Carriacou and Petit Martinique Affairs and the Carriacou

---

203 Kent, P. 76

Museum, which lists the names of the victims, and tells its own version of the origin of the mine that wreaked so much destruction in Windward.

After this incident, fishermen were warned to look out for floating mines and four more were spotted, one of which was lodged on the reef outside Windward. All of these were exploded by British navy specialists. [204] Many villagers from Windward and beyond were onlookers as the mine was exploded on the reef. A wire was run from the vessel to the mine, which was then detonated from a safe distance. Water was blown 200 feet into the air, and the blast shook the whole area.

IV: THE GRENADA VETERANS

In Grenada, all special arrangements made for the war were gradually dismantled in the months following "V.E" Day. The Local Volunteer Force was disbanded on 30th June, 1945. [205] The soldiers serving in the Caribbean Regiment, British, Canadian and other armed services, also came home as soon as transport was available, for Britain now had to repatriate the thousands of men and women who were in Britain and Europe at the end of the war. Before leaving Britain, West Indians in the armed services had the option of attending Colonial Office Rehabilitation Courses for up to a year. They could also choose to be repatriated to their own country or to the USA, Canada, or another country of their choice, or remain in Britain. Most Grenadians returned to Grenada after taking advantage of one or other of the courses offered to servicemen.

A Welfare Officer, Mrs. Florence Renwick [206], was appointed to help the approximately two hundred Grenada veterans settle back into the society. Some returning veterans could step right back into their old jobs, including those who had been civil servants or teachers. Some private sector employers had also held open the positions of employees who had enlisted.

---

204 As late as 4th September, 2008, it is said that a mine was discovered near Cane Garden in St. Vincent. It was detonated by a team from the US Navy.

205 Two years later, in 1947, the Grenada Boys Secondary School inherited the barracks at Tanteen, which were converted to classrooms.

206 Mrs. Guy Renwick

Monument to the victims of the Briquette mine that exploded in Windward Carriacou. The words read: In loving memory of those who died from the mine blast on Thursday July 6th, 1945 at 4 p.m. The mine drifted from a sinking World War II ship and came ashore at this site where local boatbuilders were constructing the MV Amber Jack. The following persons lost their lives on that fateful day: Smith Martineau Boat Owner, Patrick McLawrence Shipwright, Hyacinth Patrice Shipwright, Ronald Patrice Shipwright, Sankee Patrice Shipwright, Nicholas Roberts Shipwright, Johnny Rock Shipwright, Cicely Martineau Student, Majorie Martineau Student.

Horis Martineau,
who lost his two sisters and his father in the Carriacou Mine Disaster

GRENADIANS IN MILITARY SERVICE

Jackson Dunbar Arthur, Royal Air Force
Killed in Action

Mike Bain
Veteran of the Canadian Army,
Second Armoured Corps Reinforcement
Unit

Cosmos Cape
Veteran of the
Windward Islands Batallion

Ronald Ivan David
Veteran of the
Caribbean Regiment

Joseph Ferris, Royal Air Force
Killed in Action

Irie Francis, Veteran of the British Army.
Photo taken in Egypt before he was
demobbed.
Photo courtesy Francis family.

Cecil Harris
Veteran of the
Southern Caribbean Force

Julian Marryshow
Veteran of the
Royal Air Force

Betty Mascoll
Veteran of the British Auxiliary Territorial
Service (ATS)
Photo courtesy Kent Family

Colin Ross, Royal Air Force
Killed in Action

Elaine Moore of Grenville in ATS Uniform
Photo courtesy Elaine Moore

War Memorial Cenotaph in the Botanic Gardens, St. George's, Grenada on which are inscribed the names of all Grenadians who died in the First and Second World Wars. Photo courtesy Sonia Miller.

*They went with songs to the battle, they were young
Straight of limb, true of eyes, steady and aglow.
They were staunch to the end against odds uncounted,
They fell with their faces to the foe.
They shall not grow old, as we are left grow old:
Age shall not weary them, nor the years condemn.
At the going down of the sun and in the morning,
We shall remember them.*

From Laurenson Binyon's For the Fallen

Lest we forget.

In Memory of
Civilian OLGA MARY OTWAY DE GALE

Civilian War Dead
who died age 41
on 29 July 1944
Firewatcher; of Hanover Lodge, Regent's Park. Daughter of Mrs. de Gale, of Poyntzfield, Grenada, British West Indies, and of the late T. H. de Gale. Injured at Hanover Lodge; died same day at Middlesex Hospital.
Remembered with honour
ST. MARYLEBONE, METROPOLITAN BOROUGH

**Commemorated in perpetuity by
the Commonwealth War Graves Commission**

Courtesy Commonwealth War Graves Commission (C.W.G.C.)
(above and below)

In Memory of
Sergeant JACKSON DUNBAR ARTHUR

1394300, Royal Air Force Volunteer Reserve
who died
on 17 September 1943
Mrs M. Arthur, of Georges, Grenada, British West Indies.
Remembered with honour
OXFORD (BOTLEY) CEMETERY

**Commemorated in perpetuity by
the Commonwealth War Graves Commission**

Casualty Details

Name: ARTHUR, JACKSON DUNBAR
Initials: J D
Nationality: United Kingdom
Rank: Sergeant (W.Op./Air Gnr.)
Regiment/Service: Royal Air Force Volunteer Reserve
Date of Death: 17/09/1943
Service No: 1394300
ional information: Son of Mrs M. Arthur, of Georges, Grenada, British West Indies.
Casualty Type: Commonwealth War Dead
morial Reference: Plot I/2. Grave 69.
Cemetery: OXFORD (BOTLEY) CEMETERY

In Memory of
Sergeant JOSEPH FERRIS

1381860, 214 Sqdn., Royal Air Force Volunteer Reserve
who died
on 15 October 1942
Of Granada.
Remembered with honour
UDEN WAR CEMETERY

Commemorated in perpetuity by
the Commonwealth War Graves Commission

Casualty Details

Name: FERRIS, JOSEPH
Initials: J G
Nationality: United Kingdom
Rank: Sergeant (Air Bomber)
Regiment/Service: Royal Air Force Volunteer Reserve
Unit Text: 214 Sqdn.
Date of Death: 15/10/1942
Service No: 1381860
Additional information: Of Granada.
Casualty Type: Commonwealth War Dead
Grave/Memorial Reference: Coll. grave 4. B. 10-12.
Cemetery: UDEN WAR CEMETERY

Both courtesy Commonwealth War Graves Commission (C.W.G.C.)
(above and below)

In Memory of
Flying Officer COLIN PATRICK ROSS

132098, 49 Sqdn., Royal Air Force Volunteer Reserve
who died age 26
on 03 November 1943
Son of William Patrick and Mabel Ruth Ross, of St. Patrick's,
Grenada, British West Indies.
Remembered with honour
RHEINBERG WAR CEMETERY

Commemorated in perpetuity by
the Commonwealth War Graves Commission

Casualty Details

Name: ROSS, COLIN PATRICK
Initials: C P
Nationality: United Kingdom
Rank: Flying Officer (Air Bomber)
Regiment/Service: Royal Air Force Volunteer Reserve
Unit Text: 49 Sqdn.
Age: 26
Date of Death: 03/11/1943
Service No: 132098
Additional information: Son of William Patrick and Mabel Ruth Ross, of St. Patrick's, Grenada, British West Indies.
Casualty Type: Commonwealth War Dead
Grave/Memorial Reference: 2. II. 12.
Cemetery: RHEINBERG WAR CEMETERY

Mike Bain had enlisted in the Canadian Army in the Armoured Corps and trained as a mechanic. When it was found that he knew accounts, he was transferred to the Second Armoured Corps Reinforcement Unit as a Sergeant Clerk. He was sent to England on the *Sterling Castle* and was stationed in the army headquarters for most of the war, but he also served in North Africa, Italy, France, Belgium and Holland Germany and Canada. He was demobbed on 10th November, 1945, and returned to Grenada. When he enlisted, his employer, Geo. F. Huggins, paid his salary of fifteen shilling a week throughout the war, and his job was waiting for him on his return. Late in life he discovered that he was entitled to a veteran's pension from the Canadian Government, which was very welcome in his declining years.

Of those who had no jobs to return to, some were recruited into the police force, the prisons, the teaching service or the civil service. Others went away to study, at least three successfully completing studies in law. Some turned the training they had received as soldiers to civilian work, or went into farming or fishing to earn a living. Eighty veterans got a real break when the Government of Grenada was successful in negotiating employment for them with the Largo Oil refinery in Aruba. Others were recruited for work in Curaçao.

Soldiers were usually offered a short course when they were demobbed. Cosmos Cape, who had been trained by Lt. Henry Christopher as a wireless operator during the war, now took advantage of an additional course in radio repair, and years later augmented this with courses in the repair of televisions. These skills based on his aptitude developed in the war enabled him to make a living until he had to stop working due to his advancing years. Cecil Harris was offered a place in the peace time army, but he preferred to re-enter civilian life, and opted for a course in shoemaking. He worked as a shoemaker until his eyesight failed. After the end of the war, Ivan David was sent to St. Lucia for 6-8 months where he took a course in wireless, radio, telephone and Morse code. He was then stationed in Carriacou where he assisted Mr. Harford Mendes, eventually succeeding him as the operator of the telecommunications system for Carriacou, which comprised until fairly recently radio telephone and telegraph transmissions.

In spite of the efforts made, some veterans could find no work at all. Thankfully these were only a few.

In 1955, the Legislative Council appointed a Committee to look into the condition of the remaining veterans of World War I and World War II. The members of the Committee were: Ralph O. Williams, R.K. Douglas, T.A. Marryshow, F.L. Pierce, John R. DaBreo, A. Coomansingh and Eileen Byer. A total of 79 persons requested such help as to be given lands to cultivate, to be sold lots of Government land, to be provided with employment, to have their veteran's pension increased, to be given such a pension, or to be given general financial assistance.

Of those who were veterans of World War II, most of the complaints were about the broken promises to the veterans to give them grants of land. One or two of the veterans actually read from a booklet issued to them entitled *Demobilisation Arrangement and Facilities*, item 12 on page 11. Some veterans were unfortunate in that the funds allocated to provide them with short vocational courses were exhausted before they could obtain their promised training. The civil servants and teachers had been promised before they enlisted that upon return they would be re-appointed at the maximum salary of the scale for the positions they held on recruitment. This did not happen. Some teachers also complained that no notice had been taken of increased academic qualifications they had obtained while in service. Some veterans were now in bad health, and others had met with bad luck. T. A. Marryshow was particularly grieved about the plight of some of the ex-servicemen because, he said, he had encouraged them to fight for the British Empire and he was not satisfied that the promises made to them on enlistment had been honoured.

On 22nd September, 1955, in the same year that the Committee had finished its investigations, *Hurricane Janet* devastated Grenada. Even if the Government had wanted to, there were no available funds to devote to the implementation of the recommendations made by this committee, given the state of dire need for rehabilitation of the country, the destitution of many, and the severe injury of some of victims of the hurricane. The amelioration of the worst cases eventually fell to the Grenada Veterans Association, which had been founded in 1950. In 1957, this association became the Grenada Ex-Servicemen's Association, and finally in 2005, there was another name change to The Grenada Legion of the Royal Commission Ex-Services League (the RCEL).

Up until 2004, the remaining veterans of World War II who needed financial help received a small stipend from the Veterans Association with funds provided by the Canadian Legion. Many of the veterans suffered terrible hardship in 2004, when *Hurricane Ivan* demolished their homes. The Veterans Association did what it could to relieve their distress and since 2006, the Government of Grenada has provided a subvention of EC$ 10,000.00 to the Legion and a monthly stipend of EC$ 200.00 to destitute veterans. There is also a burial grant for the families of veterans. The Canadian Legion continues to support the Grenada Legion financially, and in other ways. The Grenada Legion raises money each year by the sale of poppies on 11th November, which is celebrated as Remembrance Day throughout the Commonwealth. [207]

But not all veterans lived in poverty and hardship. Some did well, becoming lawyers, or businessmen or estate owners. Others enjoyed employment in the services, private sector, or were gainfully self-employed. One of the characteristics of all veterans is that they seldom talk about their experiences in the war. Many were traumatised, and do not want to resurrect the terrors of war. Others think that their listeners would just not understand. Life as a soldier or airman was just too different from the life experienced in Grenada.

Due to the decline in numbers of war veterans (there were only about 19 alive in 2010), membership of the Grenada Legion has been opened to retired police, prison officers, members of the Special Services Unit (SSU), and members of the Fire Service. The Legion also has associate members from the community. In addition to looking after the welfare of its members, the Legion participates in ceremonial parades and national events and maintains links with other Veteran Associations in the British Commonwealth.

---

[207] This was formerly Armistice Day, the day on which the peace treaty ending World War I was signed in 1918.

V: POST-WAR DEVELOPMENT IN GRENADA

After the end of World War II, ex-servicemen and workers returning from Trinidad and Aruba put pressure on the Government of Grenada to provide more and better education. Those returning wanted better educational opportunities for their children. Grenadians who had been in the armed services overseas had experienced the difference that education had made to their own circumstances, because then as now, the type of job one got was decided by the level of education one had. Many had been trained in some skill during the war, or in the programmes that accompanied demobilisation, and were now better able to make a living. They had also observed the role of education in the development of society.

Increased pressure for better education also came from migrants who went to England shortly after the war to help in the reconstruction of that country. Grenadian migrants were adamant that their children in Grenada should get a good education, and sent back money to ensure this. Part of the demand for more and better education was met by the Colonial Development Fund, out of which many new primary schools, including the Grand Roy and River Sallee Government Schools, were built.

Both Britain and the United States had, before the war, tried to improve the social and economic conditions of Caribbean people. There had been riots in several Caribbean countries, deemed to be rooted in the social and economic circumstances of the majority of the population. Extensive investigations had been undertaken by the West Indian Royal Commission in 1938/39, but the war prevented the release of the Report to avoid it being used as enemy propaganda, and because there were no resources that could be spared during the war for the implementation of the recommendations. However, the Colonial Development and Welfare Organisation had been set up by the United Kingdom during the war as a result of the West Indian Royal Commission, and after the war, a number of development projects were started. Funds were also provided for other local initiatives under this organisation. Although the lion's share of these funds went to Jamaica, some benefit did come to Grenada.

The Anglo-American Caribbean Commission was also set up on 9th March, 1942, during the war

> For the purpose of encouraging and strengthening social and economic cooperation between the United States of America and its possessions in the ... Caribbean, and the United Kingdom and the British Colonies in the same area. [208]

The Commission would look into matters pertaining to agriculture, housing, health, education, social welfare, finance, economics, and related subjects. The intention was not to duplicate the British effort, but the Commission visited the British Caribbean without the benefit of having seen the Report of the West Indian Royal Commission, which had not yet been published. The Commission, therefore, produced a report with very similar conclusions to that of the West Indian Royal Commission. However, the Anglo-American Commission did do extensive research in the Caribbean during and after the war, including an analysis of the changes in wages and consumer prices in 1943. The Anglo-American Commission was also deeply involved in the anti-malarial programme, which involved the filling of swamps, including the filling of the swamp in Morne Rouge, Grenada. Apart from some alleviation of the mosquito menace, this provided space for the development of the area into Grenada's prime area for tourism.

Both these development initiatives provided the initial jobs for young Grenadians returning from universities and institutes overseas. Among them were Dr. Maurice Byer, Ray Smith and Gordon Brathwaite, who received training at the Public Health Engineering Unit and were employed in the malaria eradication programme. Leonard Berkley was also employed in the programme. His father was a tailor who had migrated with his family to Martinique. They returned to Grenada before the war began, but returned to Martinique after the war. Leonard married a Grenadian, Molly Calendar, and migrated to Jamaica.

The transformation of Grenada's labour relations was another post-war development. Labour unrest and riots, similar to the Butler Riots in Trinidad and the unrest elsewhere in the Caribbean, did not occur in Grenada due to the alleviation of population pressure and unemployment following migration to Trinidad and the ABC islands during the war, and to the tight national security. Now Grenada would explode with strikes and violence.

---

[208] Fullberg-Scholberg, P. 108

Eric Matthew Gairy had migrated to Trinidad in his youth and had spent some time involved in the activities of his fellow countryman, Tubal Uriah "Buzz" Butler. Gairy then migrated to Aruba, where his sister was already living, and was employed at the Largo/Standard Oil refinery during the war. He was sent back to Grenada when he tried to organise the workers at the refinery into a trade union. In Trinidad and Aruba, Gairy had educated himself. When he returned to Grenada, he first used his skills and made a name for himself as a community resource and activist before entering the field of trade unionism and politics. Although one might say that political upheaval and the organisation of labour was inevitable in Grenada, it was World War II that made it possible for Gairy to emerge as the person who initiated massive political and social change in Grenada.

The good achieved during the regime of Gairy must not be forgotten. Through his efforts, those who held the reins of power before him were now forced to pay their workers a living wage. But there was a lot of harm done to Grenada as well, including the breaking up of many of the plantations – derelict and profitable alike – to provide too-small plots for the Land for the Landless, without proper instruction as to how these plots should be farmed to provide a livelihood for the chosen recipients.

The war also changed Grenada's patterns of trade. Grenada and the Caribbean were to emerge with a closer alliance to the United States, and dependent on that society for many things, including continuing protection and trade. Trade was also encouraged with Canada. But in the aftermath of the war, Britain would open its doors to the large scale immigration of Caribbean people to assist in the reconstruction of the country. Traditional ties between Britain and the Caribbean, which were under threat by the new role of the USA in the Caribbean and the Americas, would again become strong.

CONCLUSION

GRENADA TRIUMPHS!

A truism oft cited is that no one really wins a war, because there are tremendous losses on both sides. This was as true for World War II as for any war. Much of the ancient built heritage of cities and towns in Germany, France, Great Britain and elsewhere in Europe was reduced to piles of rubble – the result of tons of explosives and incendiary bombs dropped on them from the air. Looters and occupying troops took care of the contents of buildings left standing. The financial cost of engaging in the war was enormous for both the Atlas and Axis countries, and left the economies of those who had engaged in the war in shambles. There can be no value put on the lives of civilians, those in the merchant marine and those in armed forces, for who can put a value on something as precious as a person's existence? The loss of human life during World War II was colossal and ripped the social fabric of many nations apart. Communities, families and individuals were sentenced to bear the trauma and scars of war for the rest of their lives. At the end of the war, events in Hiroshima and Nagasaki ushered in the era of Atomic Weaponry Stockpiles, and the subliminal fear for all persons that one day a conflict between countries will end in global annihilation.

Yet among all the catastrophes of World War II, there were quiet triumphs in places like Grenada.

Grenada had fought the war as it pertained to Grenada, and triumphed! Grenadians had not starved, or almost starved. Even though they might not have had as much to eat, or the foods that they normally ate, they had enough. They had risen to the circumstances of not having imported food by producing what they needed for home consumption, and being thankful for it, even if it was not initially the food of their choice. They had held restraint, and had patience in waiting for improvements in the various sectors of Grenadian life. They had gone about their lives in as normal a way as possible, coping with the alarms and frights of the war with courage and fortitude. Despite any grievance against Britain, many volunteered for service at home and abroad, and contributed to the war effort-by raising money, providing troops and labour, and being patient and restrained regardless of the difficulties.

They stoically bore the loss of 67 young passengers and the crew on board the *Island Queen*, and just after the war they also endured the loss of 9 more people in the Carriacou mine disaster, both of which were firmly believed to have been tragedies caused by the war. Grenadians had mourned, but had not been devastated by, the loss of four children-of-the-soil in active war service, and mourned others from Grenada and the Caribbean who sacrificed their lives in various ways during the war, many as a result of the marine warfare in the Caribbean and the Atlantic. Grenadians won their private war involving the heartbreak of loss of life, food shortages, economic hardships, transportation difficulties and occasionally fear of enemy attack.

Older Grenadians remember the excitement of the war. Some people remember the bravery of Grenadians going off to fight for Britain, and others remember the hardships. Some do not remember anything significant about this period because their lives were hardly disrupted. But in Grenada, the dangers of war were deliberately underplayed. Grenada sat for most of the war in a sea that was a hostile theatre of war, and sailors and those travelling by sea daily risked their lives to go where they had to go or do what they had to do.

Some say "God is a Grenadian". This He is, although He is at the same time the God of all nations. Nonetheless, during World War II, He looked after Grenada and Grenadians, averting from this beautiful land more catastrophe than the population could bear, supporting those affected by the traumas that did come, and assisting the entire population in the vicissitudes which descended on the population of Grenada during the 71 months of World War II.

THE END

BIBLIOGRAPHY

GOVERNMENT DOCUMENTS

Minutes of the Legislative Council of Grenada 1939, 1944 and 1945.

CO 321/386. Dispatches. Windward Islands Grenada 1939. Circular Telegram from Secretary of State for the Colonies to all Colonial Dependencies except Palestine and Trans-Jordan. d/d 15/9/1939

CO 321/386 1939 Popham to McDonald

Ordinance No 19 of 1939

Ordinance No 1 of 1919

NEWSPAPERS AND MAGAZINES

The Times (St. Vincent) August 12 & 19, 1944; Sept 2, 11 &12, 1944

The Vincentian August 12, 19 & 26, 1944; September 2,4,16 & 17, 1944 and November 11, 1944

The West Indian (Grenada) Jan 20, 1935; August 18, 1936; August 18, 1938; July 8, 1939; September 4,16,17,26, 1939; May 23, 1943, May 30, 1943; August 4,6,10,13,15,16,17,18 19,24,25,30, 1944 ; October 28 &31, 1944; November 7,9, 11, 26, 1944; December 13 and 28, 1944; January 27, 1945

The Windward Islands Annual 1955. Issues for 1956, 1957-58, 1958-59, 1959-60, 1962, 1963, 1964, 1965.

REPORTS

GOVERNMENT OF GRENADA: Report of the Commissioners appointed to enquire into the supposed loss of the Schooner *"Island Queen"* which cleared the Port of St. George, Grenada, on the 5th day of August, 1944. 31 October 1944. (Also designated as Council Paper 16th December 1944).

BOOKS AND ARTICLES

Adams, Edgar, *Linking the Golden Anchor with the Silver Chain*. Edgar Adams, St. Vincent and the Grenadines 1996.

Adams, Edgar, *People on the Move*. Edgar Adams, St. Vincent and the Grenadines 2002.

Alleyne, Warren. *Barbados at War 1939 – 1945*. Warren Alleyne, Barbados, 1999.

Anthony, Michael. *Port-of-Spain in a World at War 1939-1945. The Making of Port-of-Spain* Volume II. Second Edition. Paria Publishing Company Ltd. Trinidad. 2008.

Armstrong, W.J.C. "His Majesty's ship Diamond Rock and the German Submarine", in: *The Journal of the Barbados Museum and Historical Society*, Volume 31 November 1964.

Armstrong, W.J.C. "The Sea Devils of the Caribbean – an account of German U-boat activity in the West Indies during World War II", in: *The Journal of the Barbados Museum and Historical Society*, Volume 28 November 1960.

Anon. "Development of Tourism in Grenada", in: *Windward Islands Annual* 1962

Austin, Robin. "U Boat Warfare in the Caribbean (World War II). The sinking of the Two -Masted Schooner "Mona Marie" – Eye witness account of Captain Laurie Hassell", in: *Journal of the Barbados Museum and Historical Society* Vol. XLV Pp 124-129. 1999.

Baptiste, Fitzroy. "Exploitation of Caribbean Bauxite and Petroleum 1914 – 1945". Paper delivered at The Symposium on Caribbean Economic History. University of the West Indies, Mona Nov 7-9, 1986

Baptiste, Fitzroy. *War, Co-operation and Conflict: The European Possessions in the Caribbean 1939 – 1945*, Greenwood Press, New York, Westport, Connecticut and London 1988.

Bolleau, John "The Lady Boats", in: *Legion Magazine Canada Corner*, http://www.legion magazine.com/en/index.php/2007/01/the-*Lady*-boats

Bousquet, Ben and Colin Douglas. *West Indian Women at War – British Racism in World War II*, Lawrence and Wishart, London, 1991.

Brereton, Bridget. *A History of Modern Trinidad 1783-1962*, Heinemann, 1981.

Brizan, George. *Brave Young Grenadians — Loyal British Subjects. Our People in the First and Second World Wars.* George Brizan 2002.

Campbell, John. "Pride Before the Fall", in: *Yachting World*, March 2010 Pages 63-67.

Coard, Frederick McDermott, *Bittersweet and Spice. These things I remember,* Arthur H. Stockwell Ltd., Ilfracombe, Devon, 1970.

DeRiggs, Anthony. *Recollections of an Island Man.* Published by the author, U.S.A, 2006.

Eggleston, George T. *Orchids on the Calabash Tree.* Frederick Muller Limited. London. 1963.

Farfan, Flt. Lt. Esmond, DFC. *Five Years in World War II*. Published by the Author, Trinidad, 2010.

Francis, Irie, *As I See It: Looking Back*. Published by the author, St. George's Grenada, 2000

Fullberg-Stolberg, Claus. "The Caribbean in the Second World War", in: *General History of the Caribbean Volume V the Caribbean in the Twentieth Century*. Jointly published by Macmillan Caribbean, London and Oxford and UNESCO, Parris, 2004.

Gopaul-Maharajh, Vishnoo Franklin. *The Social Effects of the American Presence in Trinidad During the Second World War (1939-1945)* M.A. Thesis, University of the West Indies 1984.

Gentle, Eileen, *Before the Sunset*, Shoreline, Quebec, Canada, 1989

Government of Grenada, Gittens-Knight (Compiler). *The Grenada Handbook and Directory 1946*, Government Printer, Grenada, 1946.

Grenada National Museum, "Grenada in World War II 1939–1945". A Document compiled for the Exhibition held in July 2002 in Manchester, England in Association with the Commonwealth Games. 2002

Harewood, Jack, "Population Growth in Grenada in the Twentieth Century", in: *Social and Economic Studies* Vol. 15, No. 2, 1960.

Harford, Richard. *The Guava Tree*. Published by the Harford Family 2003.

Hitchens, William E. "British West Indian Airways", in: *Windward Islands Annual 1959 – 60.*

Honychurch, Lennox, *The Dominica Story — A History of the Island*. New Edition, Macmillan Education Ltd., London & Basingstoke, 1995.

Humfrey, Michael, *Portrait of a Sea Urchin*. Collins and Harvill Press, London 1979.

Hughes, Alister. "What Happened to *Island Queen*", in: *Islander* September-October 1978. Caribbean Publishing, Kingstown, St. Vincent.

Jacobs, C.M. *Joy Comes In the Morning: Elton Griffith and the Shouter Baptists.* Caribbean Historical Society, Port-of-Spain, 1996.

Jones, Claudia. *Jim Crow in Uniform*. New Age, New York 1940

Johnson, J.R. *Why Negroes Should Oppose the War*. Pioneer for the Socialist Workers party and the Young People's Socialist League (Fourth Amendment) 1939

Kay, Frances. *This is Grenada*. St. George's Grenada 1971.

Kelshall, Gaylord T.M. *The U-boat War in the Caribbean*. Originally published by Paria Publishing Co Ltd., Port-of-Spain, and Trinidad 1988. This edition published by United States Naval Institute Press, Annapolis, Maryland, U.S.A. 1994.

Kent, Edward. *Up Before Dawn*. Sail Rock Publishing, Carriacou, Grenada. 2011.

Lindsay, Jan, John Shepherd and Lloyd Lunch. "Kick 'em Jenny Submarine Volcano: A discussion of Hazards and the new Alert Level System". Paper given at the Grenada Country Conference, University of the West Indies, University Centre Grenada, January 7-9, 2002.

Lucas, C.H. "An Address to the St. Andrew's Detachment of the Grenada Contingent". N.P. N.D.

Mark, Randolph. *The History and Development of the Royal Mount Carmel Waterfalls, Grenada, West Indies*. St. Andrew's Development Organization. 1995

Marshall, Woodville K. "Entrepot and Schooner: Trade Inside the Subregion". Unpublished Paper n.d.

Metzgen, Humphrey and John Graham. *Caribbean Wars Untold – A Salute to the British West Indies*. University of the West Indies Press Jamaica, Barbados and Trinidad, 2007.

Murray, Robert N. *Lest We Forget: the Experiences of World War II Westindian Ex-Service Personnel*. Nottingham Westindian Combined Ex-Services Association in association with Hansib Publishing (Caribbean) Limited. 1996

Nunez, Elizabeth. *Prospero's Daughter*, Ballantine Books, New York, 2006

Parris, L.G.W. "The Socio-Economic Effects of the American Occupation of Trinidad (1939-1945)", Caribbean Studies Thesis. University of the West Indies, St. Augustine. n.d.

Paterson, Maurice. *So Far So Mad*, Maurice Paterson, Grenada 1994

Peebles, Margaret. *My Hughes-Steele Family from Grenada*, West Indies. Unpublished Manuscript. 2007

Redhead, Wilfred A., *A City on a Hill*, W.A. Redhead, St. George's, 1985

Special Correspondent to the Chronicle of the West India Committee. "Schooners in the Eastern Caribbean", in: *Windward Islands Annual* 1965

St. Bernard, Cosmo. "The Island Queen Disaster", in: *The Grenadian Voice* newspaper, Friday 30th July 1999.

Steele, Beverley A., "Grenada an Island State, its History and its People", in: *Caribbean Quarterly* Vol. 20 No. 1, 1974

Steele, Beverley A., "How Grenada Won World II". http://www.uwichill.edu.bb/bnccde/grenada/conference/papers/Steele.html. 2005

Tortello, Dr. Rebecca, "Gibraltar Camp – A Refuge from War", in: the *Jamaica Daily Gleaner* November 7, 2005. http://www.jamaica-gleaner.com/gleaner/20051107/news/news6.html

University of the West Indies, St. Vincent. "Some Aspects of World War II (1939 – 1945) on Life in St. Vincent" – Four short papers presented in the Local History Series. Kingstown, [197-]

Respondents

Unless otherwise specified, all interviews took place in Grenada.

Adams, Dr Edgar B.R. P.O. Box 707, St. Vincent. Interview in St. Vincent on 7th March, 2003

Anton, Beryl. Interview in Trinidad. 1st February 2006.

Arthur, Bertram "Molly". P.O. Box 264. Kingstown, St. Vincent. Interview in St. Vincent on 7th March 2003.

Azar, Roslyn, Interview on 27th November, 2009.

Bain, Arthur. Interview on 1st February, 2010.

Bain, Margaret and Mike. Interview on 2nd December, 2009.

Baptiste, Lincoln "Jack" Interview on 17[th]April, 2003

Baptiste, Jean. Interview 9th November, 2009

Barker-Hahlo, George. Interview on 26th January, 2010.

Bartholomew, Cecil. Interview on 16th February, 2010

Bishop, Alimenta. Interviewed with Maureen St. John on 23rd July, 2010

Blencowe, (Nee Fraser), Louie. 471 Sherene Terrace, London, Ontario. N6H3J4. Canada. Interview on 15th February, 2003 in Grenada.

Bonaparte, Basil and Joan. Interview on 27th May, 2011.

Brathwaite, Christine. Interview on 2nd December, 2009

Brathwaite, Godwin and Bonace. Interview on 19th January, 2010

Bristol, Carol. Interview on 14th July, 2010.

Campbell, Jenny. Interview on 9th March, 2010.

Cape, Cosmos. Interview on 15th February, 2010.

Charles, Enid. Interview on 23rd January, 2010.

Charles, Shirley. Interview on 25th January, 2010.

Compton, Patrick and Vera Compton, Interviews on 26th February, 2010 and 5th May, 2011.

Cromwell, Leopold "Leo". Interview on 27th January 2010.

Cruickshank, Arnold. Interview on 18th December, 2009.

DaBreo, Joyce. Interview on 11th March, 2010.

David, Marcella. Interview on 10th February, 2010.

David, Ronald Ivan. Interview on 6th May, 2011.

Davis, Josephine. Interview on 20th December, 2000.

DeDier, Ruby late of H.A. Blaize Street. Interview on 21st December, 2000.

DeSouza, Rosie. Interview on 19th January, 2010.

Dowe, Margaret and Reginald. Interview on 30th November, 2009.

Dumont, James. Interview on 10th March, 2010

Edwards, Cecil. Interview on 3rd February, 2010.

Ferguson, Robert. Interview on 18th February, 2010.

Francis, Adina. Interview on 17th December, 2009.

Graham, Oris Lady. Interview on 30th November, 2009

Harbin, Janice. Interview on 4th March, 2010.

Hercules, Winifred. Interview on 9th February, 2010.

Hughes, Austin. 602 E Lincoln Ave, Mount Vernon, NY 10552 – 3801, USA. Telephone interview from New York on 11th May, 2003.

Hughes, Leonard. Interview on 13th January, 2006.

Japal, Lynda. Interview on 1st March, 2010.

John, Carlyle. Interview on 28th January, 2010.

Kent, Alma. Interview on 27th November, 2009.

Kent, Edward. Interview on 24th February, 2006

Kelsick, Shirley. Interview on 20th June, 2006 in Trinidad.

Kirby, Earle. Box 752, Kingstown, St. Vincent and the Grenadines. Interview in St. Vincent on 7th March, 2003.

La Guerre, Anastasia. Interview on 10th March, 2010.

Lashley, Elinor. Interview on 26th November 2009.

MacLeish, Marjorie. Interview on 11th Feb, 2003

McIntyre, Alice. Interview on 9th December, 2009.

McIntyre, Sir Alister, 25th May, 2011. Taped interview in Jamaica conducted by Gillian Glean-Walker.

McIntrye-Campbell, Ann. Interview on 22nd January, 2010

Malins-Smith, Dennis. Interview on 11th January 2010.

Martineau, Horris. Interview on 26th February, 2010.

Mitchell, Norris. Interview on 3rd February, 2010.

Ogilvie, Harry. Interview on 2nd December 2009 and 22nd February 2010.

Otway, David. Interview on 17th February, 2010.

Palmer, Judith Lady. Interview on 18th January, 2010.

Payne, Nellie. Interview on 9th December, 2009.

Phillip, Thelma. Interviews on 14th June, 2006 and 26th August, 2010.

Pilgrim, Arthur 7 Rhoda. Interview on 2th March, 2010.

Pitt, Bertrand. Interview on 20th January, 2010.

Radix, Michael. Interview on 3rd March, 2010.

Rapier, George and Joyce. Interview on 22nd February, 2010.

Rapier, Julian. Interview on 11th December, 2009.

Renwick Hanson, Hermione. Interview on 21st January 2010

Rowley, Robert "Robby". Interview on 25th January, 2010.

St. Bernard, Cosmo. Interview on 4th February, 2010.

St. Louis, Lincoln "Brim" written personal communication. d/d 30th January 2007.

Scoon, Sir Paul and Lady Esmai. Interview on 21st January 2010.

Scott, Carmen. Interview on 28th January 2010.

Sinclair, Norma. Several written personal communication and many interviews including February 2003, in Grenada.

Slinger, Drucilla. Interview on 31st August, 2010.

Slinger, Paul. Interview on 11th March, 2010

Smith, Angela and Raymond "Ray". Interview on 15th February, 2010.

Sobers Benjamin, Geraldine. Interview on 23rd February, 2010.

Steele, Clayton. Interview on 28th January 2010.

Steele, Dunbar. Interview on 27th January, 2010.

Steele, Gordon and Valerie. Interview on 1st February, 2010.

Swapp, Gunny. Interview on 2nd March, 2010.

Sylvester, Joan. Interview on 1st December, 2009.

Taylor, A.D.R. "Sandy". Interview on 6th February 2007.

Watts, Sir John. Interview on 4th December, 2009.

Williams, Brenda. Interview on 18th January, 2010.

Williams, Joan. Interview on 26th February, 2010.

Williams, Paula and Annette Smith both née Julien. Interview on 23rd January, 2007.

U-boat icon for Chapter 3 from:

http://warandgame.files.wordpress.com/2007/10/typeixb.jpg

APPENDIX I

On the following pages are brief memorials of the victims of the Island Queen disaster, by the people who knew them then, and still remember them today. The author invites the participation of readers to send additions and corrections to this list to the author, to enhance these tributes in future editions.

A

1. Alexander, Jerome Claude. Also known as Jerome Mitchell. Jerome is remembered as pleasant and outgoing, tall, very handsome, and popular. He was the first cousin of Vera St. John, (née Alexander). He was the son of Miriam Mitchell and stepson of "Doc" Mitchell, but grew up with his godfather Jules Grant, a merchant of St. George's. Mr. Grant's establishment was on the corner of Halifax and St. John's Street, and Jerome lived with his adoptive parent on Hillsborough Street. Jerome was a good student in school, and when he left, he worked for the firm of W.E. Julien in charge of customs work. Because of his job he was familiar with the boats and their captains. Sometimes when the passenger manifest was full, or the person wishing to travel did not have the fare, or if he was short of crew, a captain would let someone travel free if he was competent to crew the vessel. On this occasion, Jerome acted as **TEMPORARY CREW** to go on the excursion to St. Vincent.

2. Alleyne, Septimus "Seppy". CREW "Seppy", who was aged about 24 – 25 at the time of the excursion, was uncle of Margaret Phillip. Originally from Florida, St. John's "Seppy" migrated to town. He and his two brothers, Cuthbert and Allan, worked on the *Island Queen*, but both brothers had chosen to stay in Trinidad after the last trip to seek other forms of employment. "Seppy" also had a sister in Trinidad, and three sisters in Grenada. When he did not return from the excursion, his partner, Alice Dumont, was left to care for their two baby daughters by herself. Her grief was intensive when one of her babies died shortly after the *Island Queen* was lost.

3. Archer, Thelma. Aged about 20, she was the eldest daughter of Frieda and DeVere Archer. She was an attractive, lively girl of light brown complexion. She was an alumnus of the Church of England High School for Girls. Thelma was to be her aunt Kypsie's chief bridesmaid.

4. Archer, Frieda. Born in St. Vincent, she was the sister of Clarice Hughes (Mrs. Russell Hughes), and Kipsey Cruickshank. Then in her fifties, she was a fantastic needlewoman. She had made all the dresses for the wedding party of her sister's wedding in St. Vincent, including the bride's wedding gown. She was the wife of DeVere Archer, who was the headmaster of the Grenada Boys Secondary School. The family had an apartment within the GBSS Hostel then on Melville Street where the National Insurance Scheme Building now stands. Frieda left her two sons, Ian who was 14 and Terrence who was 10, behind in Grenada with their father. Her 13 ½ year old daughter, Shirley, travelled by plane to St. Vincent in advance of her mother on the *Island Queen*.

5. Arthur, Michael. Michael, called "Mike", was one of six siblings: Edward, who became a vet and was known as "Bussin", Eileen Radix, Blanche Grant, Ruby La Grenade, and Jackson Dunbar "Jack" Arthur who joined the RAF and was one of the three Grenada airmen who died in Europe in active service. Michael had been a member of the Health and Strength Club along with Arnold Checkley, and once won the title of "Mr. Caribbean" for weightlifting.

B

6. Banfield, Monica. Monica was George Banfield's cousin and daughter of Jack Banfield, not to be confused with George Banfield's wife, also named Monica.

7. Bleasdille, Gertrude. Miss Bleasdille was a teacher and is remembered as a brown-skinned older passenger.

C

8. Campbell, Gordon A. (Organiser of the Excursion) Gordon Campbell was originally from Grenville, but at the time of the excursion he worked in the Public Works Department. He lived on Halifax Street. He is remembered as being an energetic, light brown man in his 30's. He liked to organise dances and other activities, and the August 1944 excursion to St. Vincent was the last of many events that took place under his auspices. The excursion to St. Vincent was proving to be his most financially successful venture.

9. Calendar, Oswald E. He is remembered as being tall, good looking, nice, polite, "quietly disposed" and respectable. He was originally from Grenville and came to St. George's to attend the GBSS. After school, he learned photography and eventually set up a photographic studio above Masanto's Store in St. George's. Masanto's survives as the three-storey building at the corner of Granby and Melville Streets. In his studio, he took photographs and processed film. He is reckoned by some to be "the best photographer that ever passed". Calendar played football for the St. George's Football Club. When the *Island Queen* disappeared he was about age 30. He left behind a wife and three children: Basil, Molly and Roland. Molly later married a land surveyor, Leonard Berkley, who was one of the professionals employed to work on the malaria eradication scheme after World War II ended. This couple eventually moved to Jamaica.

10. Checkley, Arnold. Arnold Checkley was a "big strapping guy" and a "strong man". At the time of the excursion he was in his 30's, lived on Green Street, and worked in the Public Works Department. He was a member of the Health and Strength Club and a keen body builder. He was very popular among the younger body-builders, whom he helped. He spoke with a drawl, and was light brown in complexion. His wife was a La Grenade from St. Paul's, and they had a son called Chester.

11. Clyne, George Owen Augustus. He was the son of Mr. and Mrs. H.A. Clyne of St. David's, where his father had a small estate. He was a nice young man between the ages of 20 – 25. He was a pharmacist, and operated a drug store on lower Halifax Street. This pharmacy was operated by his sister Dreda after his death.

12. Cox, Jim. This gentleman is identified as a member of the *Island Queen's* **CREW**. He is believed to be the son of a "boatman" of the same name. "Boatmen" rowed their boats for hire between Belmont, St. George's Harbour and Grand Anse. These boats had wonderful names, but were generally known as "penny boats."

D

13. de Freitas, Oona (or Patsy). She was a pretty little girl from St. Vincent who came to Grenada to attend St. Joseph's Convent. She was about 15-16.

14. DeRiggs, Lucy. Lucy was spectacularly beautiful young woman, with a lovely personality. She was loved by all, and everyone wanted to be her friend. She lived in Green Street, and worked at Everybody's Stores in the grocery department. She had just graduated from St. Joseph's Convent, and was about 19 – 20. Her sisters were Mrs. Ruby Shillingford and Mrs. Pearl Mahy.

15. DeRiggs, Hyacinth. Hyacinth was Lucy's brother, and lived on Woolwich Road with his young wife, Muriel née Layne. Hyacinth was booked on the *Providence Mark*, but when S.A. Francis would not leave his card game to board the *Island Queen* when she was ready to go, he was offered Francis' place and took it. After Hyacinth's death, Muriel never remarried, but became a well known music teacher. At one time she also ran a small school.

16. Douglas, Albert. He was the son of A. A. Douglas, the Comptroller of the Competent Authority in Grenada. He lived in St. John's Street.

17. Dumas, Violet. She is believed to be one of the Trinidadians who came up from that island on the *Island Queen* to go on to St. Vincent. The *Island Queen* had brought up a few passengers from Trinidad who travelled with the Grenada/St. Vincent excursion.

E

18. Evans, Jean. Jean was the daughter of David Evans and Ethel Lumsden of Victoria. She was 19 years old, and a graduate of Church of England High School for Girls. She boarded at Mrs. Rapier's house on Lucas Street, one of the homes-away-from-home for schoolchildren who lived too far from the school to return home every day. Mrs. Rapier also provided lunch for children who lived near enough to town to commute, but too far away to go home for lunch. Jean had just got her first job at the *West Indian Newspaper*. She was the sister of Marjorie Rapier Cameron and cousin to Glyn Evans and Esmai Lumsden. Honey Rapier was one of her good friends. She was a pretty, fun-loving, happy-go-lucky person. She was very light skinned, and had lovely, long reddish-brown hair which she wore in an upsweep – the popular style of the day. Jean was booked on the *Providence Mark* but her friends on the *Island Queen* persuaded her to join them. Her suitcase travelled on the *Providence Mark*, and on its return to Grenada was handed over to her sister Marjorie.

F

19. Fletcher, Irva. Irva was a very pretty woman in her early 30's. She was originally from Victoria and sister to Claire Fletcher. Eva and Edris Fletcher were her cousins. She was a civil servant.

20 & 21. Fraser, Jean and Patricia "Patsy". Jean and Patsy Fraser, aged 16 and 14 respectively, were the daughters of the Hon. Alex Fraser and Mrs. Erene Fraser from St. Vincent. They were pupils attending the St Joseph's Convent in St. George's and were going home on the *Island Queen* for the holidays. They were the youngest of seven daughters: Louie, Agnes, Eileen, Laurie, Oonah, Jean and Patricia. Jean had a beautiful voice, and used to sing in the school choir. She also sang solos at celebrations for Empire Day, which were held in the Market Square. On these occasions, school children marched from their schools to the square, and back after the ceremonies. When they returned to school they were given a bun and a sweet drink. Jean was a friend of Esmai Lumsden. The loss of Jean and Patsy profoundly affected the Fraser family, the father dying soon after from a stroke, their mother never herself again, and the sisters living with the grief.

G

22. George, Sheila. Sheila George lived in one of the bungalows on the Carenage. She was in her late 20's, and a Civil Servant. She was the first girl with very dark skin to win a scholarship to secondary school, in the days when there was only one scholarship for girls and one for boys. She attended the Church of England High School for Girls. This clever student was also a Guider.

23. Gifford, Derrick. Sometimes Gifford worked as a stevedore, and at other times he was a member of the **CREW** for various vessels.

24. Glasgow, Justina. She was born in St. Vincent, and worked for Mrs. Archer. "Tina" was going to help with the wedding, and was also in charge of the Slinger children.

25 & 26. Gomas, Rita and Sylvia. Rita and Sylvia were sisters, and were both in their late teens and very popular. They were both still students at St. Joseph's Convent. They lived on St. John's Street, opposite the Day Nursery. They were cousins to Jack Baptiste. There were four other sisters — Clythe, Erma, Ena and Eileen and a brother named Eric. All of the sisters were very musical, and were often in demand for renditions on the piano, mandolin, ukulele or guitar. They also sang. Ena is also credited as being the first woman to own her own business in Grenada, and the first person to develop an industry to support the fledgling tourist trade. She taught the country people how to prepare the *lapit* for the manufacture of straw goods, and then purchased what they produced. Ena would eventually purchase a part of the Supplies Stores on Granby Street, where she already had her shop which sold straw work, local articles and tourist souvenirs. Ena also decorated floats for Grenada's Carnival. Cecil Bartholomew was the child of Eileen Gomas and her husband Dowlin A, Bartholomew. He was raised by Ena after his mother died.

H

27. Howard, Henry. This young man was a teacher at St. Paul's Model School. He lived on Park Lane.

28. Howell, Lennox. He was a St. Lucian living in Grenada, and was a friend of Jack Baptiste. He was a body-builder and used to lift weights. Jack and Lennox had packed their food for the *Island Queen* excursion together and when Jack transferred to the *Providence Mark*, they sat together on the wharf and divided the food.

29. Hughes, Clarice -Née Cruickshank. Remembered as a very nice woman in her 40s, Mrs. Hughes was going to her sister's wedding in St. Vincent. Her husband was Russell Hughes, who worked at Geo F. Huggins. Russell and Clarice had one daughter, Dawn.

30. Hughes, Ian. Ian was about 17 at the time of the disaster. He was the eldest son of Earle Hughes, a member of Grenada's Executive Council, and a member of a number of Boards including the Board of Directors of the firm of George F. Huggins, the Grenada Co-operative Bank, the Bus Company and the Grenada Building and Loan Association. Earle Hughes was employed by the firm of George F. Huggins. The family lived in Church Street. Ian was a good friend of May Donovan, sister of Nellie Donovan (Payne), and also a good friend and sailing partner of Paula and Annette Julien and George "Porgy" Rapier. Ian was always getting small injuries from adventurous exploits, and was "patched up" regularly by Porgy's mother who was a nurse. Described by many respondents as a "very nice young man", he had just completed his studies at the GBSS and was preparing to go to McGill University. Many believed that he would eventually get engaged to his girlfriend, Honey Rapier.

31. Husbands, Huille. Huille had been a prefect at GBSS, and at this time worked in the Customs. Some remember him dressed smartly in his white drill uniform. He was the elder son of a "gentleman" tailor, Mr. C.A. Husbands. His father was also an expert in the cutting of meat and was in great demand whenever there was an animal to be slaughtered for the market. Huille's father went to such engagements in great style, complete with coat and tie. He was the elder brother of Colin Husbands, and brother to Joyce. Huille was nephew of John, Walter and Gittens Knight, who were his mother's brothers. His family owned and lived in a building near the top of Halifax Street where one of the Purcell's buildings now stands. Huille was the only member of the *All Blacks Football Team* who chose to travel on the *Island Queen* rather than with the rest of the club on the *Providence Mark*. He wanted to travel with Thelma Archer, who was his serious girlfriend, and his team mates' urgings could not shake this resolve.

I, J

32. James, Eileen. At the time of the disappearance of the *Island Queen*, Eileen was still a student at the St. Joseph's Convent, and lived on St. John's Street.

33. Johnson, Sinclair.

34. Joseph, Lennox. CREW.

K

35. King, Joseph. Joseph King was a Trinidadian who used to come to Grenada quite often. He was a friend of Chykra Salhab.

36. Knight, Ivor. He was the eldest son of John Knight who at the time was secretary to the Windward Islands Government and worked at Government House in St. George's. John Knight later became the head of the Postal Department in Grenada. Derek Knight was Ivor's younger brother. Ivor and Derek were first cousin to Huille Husbands. Ivor was

Godson to Lillian Roberts of Hyde Park in the Villa, the organist at the Presbyterian Church. He is described as being "very nice", and in his early 20's at the time. He had just completed his secondary schooling at Harrison's College in Barbados, and had got a scholarship to study Medicine. He told many of his close friends that he was going on the *Island Queen* to have a "last fling" before travelling to England to commence his studies. His scholarship was awarded to Hilda Gibbs (Bynoe) when the *Island Queen* did not return.

L

37. Linck, Donald. Donald Linck was from Mount Airy, and at this time was between 20 and 22. He was a good looking brown skinned young man, who had just left school and had been taken on at the Grand Anse Sugar Factory by Humphrey Parris. He lived with the Parris Family at the Grand Anse Estate House which was situated on the hill overlooking the works. He had many friends, including the Buxo and Buckmire families especially Maurice "Sonny" Buckmire. He was a schoolmate of Ray Smith, and taught him to ride a bicycle. He is remembered for his own bicycle riding. He was a past student of Fletcher's School and GBSS.

38. Lindsay, Rupert. CREW.

M

39. Marryshow, Hans. He was the tall, handsome son of T.A. Marryshow and Edna Gittens. He lived with his mother and siblings — Basil, Glyn, Sheila and Louise — in a house located half-way down Briggs Alley. Basil became a doctor in the United States of America and Glyn made a career in Public Health. These of Marryshow's children were also half siblings to Ivor, York, Selwyn and Bert Marryshow and Hermione and Curtis Charles. Before the excursion, Hans had been called by his father and instructed to look after his sisters on this journey. Ivor, his half-brother, travelled on the *Providence Mark*. Hans, in his twenties at the time of his death, was a teacher at the Woburn Methodist School. He was a classmate of Ivan David when he was a pupil at the Wesley Hall School, and Ivan remembers that Hans was the brightest in the form.

40 & 41. Marryshow, Louise and Sheila. Louise and her sister Sheila are described as "gorgeous girls". They had attended Wesley Hall School, and were classmates and friends of Thelma Knight (Phillip). Louise was in her early 20s and ran a small school at home. Sheila was younger than Louise, and at the time of the disaster, worked at the Square Deal

Store owned by E.D.B. Thomas which was then located at the Corner of Granby and Melville Streets. The death of three of his beloved children "took the heart out of their father", and T.A. Marryshow was observed never to be the same again. This was also true of their mother, Edna Gittens.

42. Minors, Merle Merle Minors was a Form 4 student at St. Joseph's Convent. She was travelling home to St. Vincent accompanied by her aunt Bernice Swapp, with whom she lived on Tyrrel Street when she was in Grenada to attend school. She was the daughter of Rupert and Madeline Minors of St. Vincent. Nicknamed "Topsy", Merle was 14 years old at the time of her death, and the youngest of 8 children. Her siblings were Ivy, Angela, Jack, Iris, Dan, Nesta and Billy. Merle is remembered as a brilliant girl, usually coming first in her class at the end of each school year. Two of her close school friends were Esmai Lumsden (Lady Scoon) and Judith Parke (Lady Palmer). William Otway was her cousin.

43. Mitchell, Lewis. CREW.

44. Moore (Mrs. Phillip), Perle "Jappy". Jappy was a very pretty girl from Grenville who now lived in St. George's and worked at the Bata Stores. She was the eldest of the 7 daughters of Preston Moore, the manager of Everybody's Stores. Her sisters were Jocelyn, Eileen (called "Chickie"), Elaine, Ailsa, Phyllis and Katey. She also had a brother called Winston. The Moores were a bright and talented family, and all the sisters were talented singers. "Jappy" and her sisters all attended the Church of England High School for Girls, and lived with their aunt, Florrie Moe. When she went to work, Jappy used to take lunch at Mrs. Commissiong's boarding house located on Halifax Street above Everybody's Store. Jappy married George Phillip, one of Grenada's top tennis players. George Phillip was at the time of the disappearance of the *Island Queen* in the Caribbean Regiment and was stationed in Italy and Egypt. Jappy and George had an infant son, Michael. After the *Island Queen* went missing, Irene Moe, Jappy's first cousin, took Michael, and brought him up like her own son. On his discharge from the Army, George Phillip went to Curaçao where he got a job, but remained devoted to Michael, supporting him in every way. Michael attended the University of the West Indies, and later had a very successful career in the United States.

O, P

45. Parris, Humphrey N. Humphrey Parris was the husband of Madeline, one of Chykra Salhab's younger sisters. They had three children, Noreen, Joan and James. Prior to leaving Barbados, he had enlisted and served in the First World War – "The Great War for Civilization" –

in the West Indian Regiment. Originally from Barbados, he was well connected, and was known as a gentleman and nice person. His friends included Albert Gomes, Alexander Bustamante, Norman Manley and T.A. Marryshow. He worked for Grand Anse Estates, and lived in the Great House on the property. Many remember that he loved to take his wife for drives in a horse and buggy. He himself rode a horse around the estate. Joan Williams, his daughter, remembers him playing with his children, especially "horsie", with them riding on his back. He was nevertheless, a strict disciplinarian. Humphrey Parris needed a part for a piece of machinery at the sugar factory which he thought he might get in St. Vincent. He took advantage of the excursion to travel to St. Vincent, taking Donald Linck, his young assistant, with him.

46. Paterson, Claire. Claire was in her early twenties. She was sister to Dorothy Paterson, who later married John Watts. She worked at Granby Stores as a cashier. She had a beautiful voice, and had been asked to sing at the Cruickshank wedding in St. Vincent. She is remembered as pretty and vivacious, and had the habit of carrying a pencil on a string which she would twirl to accompany whatever she was doing, including walking down the street. Jenny Campbell's last memory of Claire dates to the year before her death when Claire was just about to leave the Church of England High School. Jenny remembers seeing her up on a ladder in the school library attending to the books.

47. Paterson, Lucy. Lucy was the daughter of Fred Paterson from Carriacou and Molly Hill. She lived with her mother in Sauteurs where she taught at the St. Patrick's Roman Catholic School. She was cousin to Claire Paterson and Dorothy Watts, and half sister to Reginald and Joslyn St. Bernard. She is remembered as plump, very nice, a loving person and a caring teacher.

48. Peters, Ruby

Q, R

49. Rapier, Helena "Honey". Honey was nineteen years old and a graduate of the Church of England High School for Girls. She was also a very nice girl with a beautiful smile. She is described as being stunning, gorgeous, vivacious and "such a lot of fun". She had been in the Brownies as a child, and was given the name Honey because she was as sweet as honey, and also because she had honey-coloured skin. She was originally from Grenville where her father, Cecil Rapier, had a big business. Her mother had a house on Lucas Street where Honey lived. Honey was sister to Meryl St. Bernard and cousin to Fair, Thelma, Barry, and Erva Rapier. She was friendly with Jean Evans, who boarded with her mother. She was the girlfriend of Ian Hughes, and many believed that they would eventually become engaged. The day of the excursion, she was seen running down Lucas Street to catch the boat because she thought she was late, not knowing that the departure of the *Island Queen* was to be much delayed. Her parents never recovered from the loss of their sweet daughter.

50. Richards, Lester. He was married to Joan, and lived in Trinidad and Tobago.

51. Richards, Joan. She was from St. Vincent, and lived in Trinidad and Tobago where she was a teacher at Bishop Anstey High School. The couple were going to St. Vincent on holiday.

52. Rowley, Redvers. He was the brother of Louise and Doris Jane Rowley. Their mother was a nurse and their father was a schooner captain. Redvers had served as District Officer for Carriacou which included duties as Harbour Master and in the Customs and Excise. At the time of the excursion, he had just recently come back to Grenada with his wife and three children, on promotion to the Customs Department. He decided on impulse to go on the excursion just for the fun of it. He was a dark, clever, well-spoken and popular man, with a charismatic personality. He was good at calculating, and was often asked to tally at horse-racing events. He had a "special walk", which was a rolling gait typical of sailors. At the time of Redvers' death, he was 39 years old. For the rest of her life, Iris, Redvers' wife, kept his picture close beside her always. She died in January 2011 at the venerable age of 103..

S

53. Sadowick, Elias. It is believed that this passenger was a Trinidadian and friend of Chykra.

54. Salhab, Chykra. CAPTAIN. Affectionately called "Uncle Chickie" by his nieces, nephews and youngsters close to him, Salhab was about 60 years old at the time of the tragedy. He was born in Syria, and was brought to Grenada at age 4 by his parents. Although he grew up in Grenada, he still spoke accented English. He was the eldest of eight children — Chykra, Louis, André, Matilda, Isabel, Madeline, Zahia and Rosa.

After his father died, Chykra assumed responsibility for the extended Salhab family which included his mother Dora who still presided over the household, various "live-in" nieces and nephews, and hosts of temporary family children and grandchildren who, although not resident, thought of themselves as rightfully belonging in "Mooma's" family. A year before the loss of the *Island Queen*, Chykra married Olga Smith, a Grenadian, and moved with his wife and their two children — Cynthia and Gordon- to a house in Morne Jaloux. After Chykra married, he still managed the Salhab household and family complex on Scott Street, the lower apartments of which complex were utilised for the many enterprises of the Salhab brothers. Chykra had two other children, Jean and Willie, who he continued to support and to care for. Until their mother died, Jean and Willie lived with her in St. George's, coming to live with "Mooma" in Scott Street after they were orphaned.

Chykra was a particularly good and hard-working businessman, involved in many enterprises besides sailing, all of which prospered. With his brothers, André and Louis, Chykra operated a shipping agency, a garage, a bicycle shop which held the agency for Runwell bicycles, and a mechanics shop. The shop which sold the bicycles along with other related items was called the *Hole in the Wall*. The family also had trucks which were parked in a garage under the Hayling's house in Scott Street. Louis, a mechanic, drove the Salhab trucks and worked on the engines when necessary. Chykra also operated a taxi, and ran a luxurious ten-room hotel in Morne Rouge, where Mount Cinnamon now stands. Many Trinidadians came to stay there and enjoy the first class accommodations, which included gold plated tableware.

One of Chykra's sisters, Madeline, married Humphrey Parris who was a passenger on the *Island Queen*. Another sister, Isabel, called "Zabelle", married Osborne Steele, a mechanic like Louis. Osborne worked with Louis Salhab driving and servicing the trucks and working on the engine of the *Island Queen*. Louis and Osborne also frequently crewed on the *Island Queen* with Osborne sometimes skippering. Louis Salhab had 4 children, Louie, Louis, Agnes and George. The family lived in *Bluggoe Cottage* on Williamson Road.

André Salhab, Chykra's other brother, enjoyed his life to the full. He was a keen spear fisherman, and taught others how to fish using this method. He was a wonderful swimmer. He particularly enjoyed playing carnival, and many remember him engaged in making carnival costumes, and playing 'Mas. He also liked Christmas and New Year's, and celebrated these in great style. At these times his house at the corner of Glean's and Springs Roads was ablaze with lights, and a sight to behold. Elinor, Dennis and Phyllis (in St. Vincent) were his children. Elinor shared her father's love of Carnival, and was herself one of Grenada's Carnival Queens.

Chykra was known as a responsible person, who never drank to excess. He was jocular, kind-hearted and helpful. He was also an experienced sailor, sailing for pleasure with Colin McIntyre. Chykra had previously owned a motor yacht called *Cassandra*, which had been completely destroyed in a storm. Now he operated the *Island Queen* which usually plied between Grenada and Trinidad transporting people and goods.

The entire Salhab family was known for their generosity to the community. Chykra's mother, Dora, used to feed everybody who needed a meal, and bake goodies as gifts for nearly everyone in Scott Street. She is described "a very good woman". "Zabelle" in the same tradition, made clothes for those who needed. Rosie DeSouza and her sisters lived near to the Salhabs and were very grateful to Zabelle for the clothes she made for them.

55 & 56. St. Bernard, Reginald and Joslyn. Reginald and Joslyn were a brother and sister originally from Sauteurs, St. Patrick's. Their father was a very senior member of the Catholic community in Sauteurs, and their mother was Miss Molly Hill who lived in upper High Street, near McDonald College in Sauteurs. Reginald, Joslyn, and their half-sister Lucy Paterson lived with their mother Molly. Reginald was 35 or 36 at the time of the excursion. He was of average height; light complexioned and handsome, with curly hair. For many years he was a much loved teacher at the St. Patrick's R.C. School. He was very active as the scoutmaster of the 2nd St. Patrick's 18th Grenada Scout Troup. He often took the scouts to camp. Reginald left teaching to spend some years in Curaçao. On his return he got a job in the civil service. He was engaged to Thelma Steele who was a sister to Osborne Steele and one of Magistrate Henry Steele's children. Joslyn was in her late twenties, or early thirties at the time of the excursion. She was a seamstress who loved to sew for her mother, her sisters and herself. She is remembered as being stylish, slim, tall and beautiful.

57. Scoon, Doreen. Doreen was from Grenville. She was the sister of Dorabella Shears, and cousin to Cicely Peters. She was a distant cousin to Layinka Scoon and Sir Paul Scoon. She was brown-skinned, beautiful and "well figured," although of slight stature. Doreen was a popular, outgoing, friendly girl with a soft, nice manner. She loved to dance, and was a member of the Physical Culture Club in Grenville founded by Gascoigne Blaize. She was also an excellent swimmer and played basketball. Employed at the firm of T. Noble Smith in St. George's, at the time of the excursion she boarded in town. Some days before the excursion was due to leave, she paid a visit to her mother in Grenville to let her mother know that she was going on the excursion. At the time of her death she was between 22 and 24 years old.

58. Scoon, Layinka Dorothy. Layinka was just 19 years old at the time of the tragedy. She is remembered as being a very quiet, very pleasant person. She was first cousin to Sir Paul Scoon, and a cousin also to Enid Charles. Layinka's sister, Winifred (Hercules), was 9 at the time of the excursion. She remembers that Layinka had just passed her Senior Cambridge Examinations with the help of de Vere Archer, who coached her in mathematics. Layinka had just started working at the Post Office, and with money from one of her first pay packets, had bought herself a pair of multicoloured platform shoes. Winifred remembers admiring them, and determining that she, too, would get a pair like her sister's when she grew up.

Layinka is an African name, for Layinka had been taken to Africa as a baby, and returned to Grenada when she was 12. Her father was one of the several Grenadians who accepted contracts to develop the cocoa plantations or to work on building the railway system in Nigeria. Some of the other Grenadians who went were John Louis Buckmire, Conway Steele, Willie Hagley, Espinoso Clyne and Leopold Cromwell who was the brother of Dunstan Cromwell.

Layinka was on her way to St. Vincent to visit Mrs. Rose Jennings, who had been in Africa with her parents. Marie Scoon, her mother had suggested that she travel with yet another family friend, Miller David, who had a passage on the *Providence Mark*. However, Layinka preferred to go on the *Island Queen* with all the other young people and Mrs. Scoon eventually gave in to her daughter's choice of boat. Marie was at her window when the *Island Queen* sailed past the Esplanade. She could see the young people on board already having a good time. Winifred had been sent to Mrs. McEwen, a family friend, in Grenville for the school holidays. When it was apparent that something had gone terribly wrong with the *Island Queen*, she was called back to town by her mother without knowing why. When she got home, her mother greeted her with the bad news.

Marie Scoon ran Scoon's Guest House in St. John's Street which catered for schoolchildren from the country who needed accommodation in town. During the school holidays, Scoon's Guest House was the first choice for accommodating sports teams and their supporters coming to play in Grenada against Grenada teams.

When improved transportation made it unnecessary for school children to board in town, Scoon's became a regular guest house, accommodating

18 visitors, or locals who needed a place to live in town. When Mrs. Scoon died in 1959, Winifred continued to run the establishment with the help of Mrs. Eileen McIntyre. Eventually the guest house was closed, but Winifred continued to live in the building. In 1998 the building was sold. Grensave and other offices now occupy this three storey building.

59. Scott, Irie. Irie Scott was regular member of the crew of the *Island Queen*. He was the son of John Scott from L'Esterre, Carriacou and at the time the vessel was lost, he was in his 30s. Apart from the loss to his family, he left bereft a girlfriend and an infant son. The death of Irie's baby shortly after the *Island Queen* was lost compounded the already unbearable tragedy for those who loved him. Irie Scott was the only person on the schooner who was a native of Carriacou.

60 & 61. Slinger, Dawn & Denise. CREW. Denise and Dawn were aged 12 and 14 when they boarded the *Island Queen* bound for St. Vincent. They were the young daughters of Dr. Evelyn Slinger and Ena née Richards. Ena Slinger was a Vincentian. The entire family was invited to the Cruickshank wedding, but only the children would go since Dr. Slinger was away. Dr. Slinger was the surgeon at the Hospital in St. George's and lived with his family at the Surgeon's house called *Rathdune* near the hospital. Dr. Slinger's previous posting was the medical officer in Grenville. The Slingers had then lived in the house rented for the Medical Officer on the Boulogne Estate.

Dawn had short brown hair, and is remembered to be a quieter child than her sister Denise. She greatly resembled her cousin Valerie Renwick (Steele). She enjoyed her annual visits to St. Vincent, and loved to dance to pop music. Denise had long blond hair, and was very pretty. She also loved to dance to the popular dance tunes of the day.

Dawn and Denise were pupils of the Church of England High School for Girls. Classmates remember these children as nice, regular, outgoing girls. Both Denise and Dawn were music pupils of Mrs. Evelyn Pilgrim, who lived at that time, on Lucas Street. Arthur Pilgrim, Evelyn's son, remembers these two young ladies coming for their music lessons.

The disappearance of the *Island Queen* with these two girls had a traumatising effect on the pupils of the Church of England High School, especially those in the same forms as the girls. They had to look every day at the empty desks and wonder where their classmates were. The children told each other all sorts of tales and for a very long time expected to hear that Dawn and Denise had been found alive in some part of the world. Many of the people who were their schoolmates still mourn and pray for them.

The children were also related to Christine Gun-Munro (Brathwaite) and Dr. Slinger was Christine's godfather.

62. Stroude, Godwin. Godwin was about 20 – 22 and worked as a store clerk at the firm of T. Noble Smith. He hailed from LaBorie. His cousin Alfonse worked in the Government Printing Office. Alfonse travelled on the *Providence Mark*.

63. Steele, Estelle. Estelle was a very attractive daughter of Alfred "Ten" Steele, the manager of the Antilles Hotel. She was a nurse, and lived on Woolwich Road with her mother.

64. Steele, Thelma. At the time of her death, Thelma was about 25 years old. She worked in the Government Treasury. Thelma was the youngest child of Henry Steele with "Miss Sandra", and lived with her mother and siblings: Russell, Osborne, and Beatrice, (called "B"), in Scott Street. She was one of the many persons who "had the run" of the Salhab house on Scott Street because Osborne, Thelma's brother, married Chykra Salhab's sister Esabelle ("Zabelle"). Thelma was therefore Chykra's sister-in-law. Thelma was engaged to Reggie St. Bernard. Her sister "B" was a very tiny person who wore rimless spectacles, had a lovely singing voice, sang in the Church of England Choir, and taught in the Sunday school.

65. Swapp, Bernice. The Swapp family was originally from St. Vincent, the original progenitor being brought from St. Vincent for a special job on the Brandon Hall Estate. Bernice was accompanying her niece, Merle Minors, home for the holidays and expected to use the trip for a little vacation for herself. Remembered as "such a lovely woman" she lived close to Muriel Fitt on Hillsborough Street, and was her friend. She worked at Everybody's. The story is told that as Bernice approached the *Island Queen*, she slipped, and since this was said to be unlucky, she turned to board the *Providence Mark* instead. People laughed at her believing in such nonsense, and she reacted by boarding the *Island Queen* as scheduled.

T, U, V, W

66. Williams, Edith. Edith Williams lived at The Rock, St. George's where Bobby's Tyre Mart now stands. Her elder sister, Emily Williams Phillips also lived at The Rock. Her niece, Grace, married Ira Young of St. Vincent in 1943, and now had an infant son, Robert, who in August 1944 was 4 months old. Edith decided to take advantage of the trip on the *Island Queen* to visit her sister and brand new nephew. Grace was godmother to Marcelle Ross (née Heywood). Her godmother wanted to see Marcelle, who was about 5 years old at the time, and asked that her sister bring her over to St. Vincent. Plans were proceeding with Marcelle's parents for the journey, but somehow they were not completed in time, and Marcelle did not make the journey. Had she gone, she would have been the *Island Queen's* youngest victim.

67. Williamson, Ernest. Ernest and George were the sons of Arnold Williamson, Chairman of the St. George's District Board, and Constance Williamson. Both brothers were graduates of GBSS. They were brothers of Alice McIntyre and had two other sisters. Ernest was nicknamed "Pram". The Williamson family lived above their father's garage and petrol station on Young Street. Ernest had just returned from the United States as a qualified mechanic, and was to take over the part of his father's business that required this expertise, while his brother, George, would take over the accounting and management side of the business. Both brothers were to travel on the *Providence Mark*, but on impulse Ernest jumped over to the *Island Queen* at the last moment, leaving George on the *Providence Mark* will all of his possessions.[207]

---

[207] The documents cite the number of passengers and crew as 67. However, the boat was possibly carrying Ernest Williamson and Mike as supernumerary, raising the number to at least 69. However, I have been unable so far to identify any other persons who travelled on the *Island Queen*. I have also been able to identify only 9 crew including Chykra, which means that there were one or two other persons who travelled on the *Island Queen* acting as crew on that fateful journey.

APPENDIX II

On these pages are brief memorials of those who travelled on the Providence Mark, and so escaped death on the occasion of the fateful excursion from Grenada to St. Vincent on 4-5 August, 1944. The memorials were compiled from the memories of the people who knew them then, and still remember them today. The author invites reader to bring any additions and corrections to the attention of the author, so that these can re rectified in future editions.

A, B

1. Banfield, George. He was an employee of the firm of Jonas Browne and Hubbard, first in the Lumber Yard, but eventually working himself up in the firm to retire as a joint Managing Director of that firm. George Banfield was Grace Steele's father. He was married twice – first to Altehea Cochrane and then to Monica Vincent. He is remembered as a very nice man, and as he aged, he had a wonderful head of white hair.

2. Baptiste, Lincoln "Jack". Jack was 20 years old in 1944. He had gone into jewellery making, specialising in wooden and tortoise shell broaches and ear-rings. He had contracted rheumatic fever as a teenager, which curtailed his participation in sports, but he was a non-playing member of the All Blacks Football Club. He was cousin and neighbour to the Gomas girls, and they all lived on St. John's Street. Jack lived to be one of Grenada's well-known businessmen, owning with his wife Jean the Seachange Bookshop, and a meat processing plant.

3. Baptiste, Verna. She is remembered as being a small slim lady who would have been in her 20's at the time of the excursion. She was a seamstress.

4. Brathwaite, Roslyn. Roslyn was the sister of Gordon and Edward Alexander "Neddie" Brathwaite. She married Solomon Azar, and assisted him in the running of a successful restaurant and later a bakery and patisserie in St. George's. She is the mother of four daughters: Katy, Sandra, Nanette and Carolyn.

5. Buxo, Lillian. Oswald and herself were siblings of Flo Buxo.

6. Buxo, Oswald. Oswald was employed to the Ministry of Education as a visual education officer. He married Adelaide Banfield, first cousin to George and Monica Banfield. Anthony "Tony" Buxo, former owner of Grenadian Optical, was his adopted son.

C

7 & 8. Charles, Beryl and Rosemary. Beryl and Rosemary were sisters to Jocelyn, Eunice, Bert, Alistair and Ivan. In the 1960s and 70s, the family ran one of the most modern stores in St. Georges, called Charles of Grenada. Jocelyn, Eunice, Beryl, Rosemary, Bert, Alistair and Ivan were first cousins to Roslyn Brathwaite. Rosemary, nicknamed "Sylvie" is best known as one of St. George's artists and florists. She was also an outstanding churchwoman, and an elder of the Presbyterian Church. Bert called "Bertie" was a doctor in St. Vincent.

9. Charles, Curtis. Nicknamed "Pappy", he was Hermione Charles' brother, and half-brother to York Marryshow. He was a cousin of Bert, Alistair, Ivan, Jocelyn, Eunice, Rosemary and Beryl Charles. He is remembered as a tall attractive man, and a very charming person. His son at one time managed the Point Salines International Airport.

D

10. DeRiggs, John. This was the handsome brother of Lucy de Riggs. At the time of the disaster, he was 26 years old and lived with his sister, Lucy, in Green Street. Both worked at Everybody's Stores. Sometime after World War II ended, John migrated to the United States.

11. David, Miller. From Victoria himself, he owned and operated a bus on the Victoria to St. George's run. He was a friend of Layinka' Scoon's family. Marie Scoon had hoped that Layinka would travel with him to St. Vincent on the *Providence Mark*, but Laynika pleaded to be allowed to travel with her friends on the *Island Queen*.

12. Donovan, Selby. Selby was the brother of Nellie Donovan (Payne) and her sisters Daisy, Ethel May, Jessie and brother called Vince. Selby, 22 years old at the time, was a popular person from a popular family. He was handsome, cheerful and happy-go-lucky person. He was a member of the All Blacks Football Club. Originally booked to travel on the *Island Queen*, he transferred to the *Providence Mark* at the last minute. Later on in life he migrated to the United States of America where he married, and worked in the Demolition Division of the Building Department of New York City.

13. Dowe, Edward Flemming. Nicknamed "Fello", Edward was 35 years old at the time of the excursion. He was born at *The Rosary* on Tyrrel Street. He married Dora Smith, and made his home in Morne Jaloux. A member of the All Blacks Football Club, he was employed with Geo. F. Huggins until his retirement in 1975, and is described as a "responsible fellow". He was the father of Reginald "Reggie" Dowe and Joan Dowe-Janes. He was the grandfather of Colin Dowe.

E, F, G

14. Gittens, Eversley W. This gentleman was a general merchant and later a prominent entrepreneur. He was the father of Dr. Bernard Gittens. Really determined to travel on the *Island Queen*, he was sorely disappointed when he found he had to travel on the *Providence Mark*, or not travel at all.

H

15. Hall, Mark. CAPTAIN of the *Providence Mark*. Mark Hall was a man who was both a "respectable" man and a man highly respected. From a very poor family in St. Patrick's, Mark Hall lived and operated from Grenville at the time of the *Island Queen* tragedy. From boyhood, he greatly loved the sea and his favourite place was on the jetty in Sauteurs. It was from this spot that Mark Hall was recruited for his first job as cabin boy on a small freighter.

After working at this for a few years, he left the boat in Alabama, where he worked at dishwashing and other menial jobs. Experiencing the racial prejudice that was brutal in the southern United States at this time, he stayed only long enough to save a little money before returning home to Grenada to build and captain boats. While he was in Alabama he had several gold teeth put in, which are remembered as part of his persona. He is remembered as a very dark, short, strapping, stocky, muscular man with a pleasant face. He was helpful and kind, gruff but gentle.

He loved children, and would allow them to play around his boat. Each year as Christmas approached, he would shop for toys and balloons on stop-overs in Trinidad and elsewhere. He would then decide on a day, "take a car," and go around distributing the toys to the children in St. Andrews and St. Patrick's. Hall had previously owned the popular sloops The *Mary Rose* and the *May I Pick*. The *May I Pick* had an unfortunate end when in very bad weather she got stuck on the reef that encloses Grenville Harbour where she started to break up. She was carrying many passengers, and some lives were lost, mainly because of panic, and the fact that many could not swim. The *May I Pick* was replaced with the *Providence Mark*. Mark Hall also owned the *Lady Kelvin*. This boat sank in 1957 (after he had died) when it was accidentally rammed by a tanker, killing his brother-in-law and six other people. After the ill-fated excursion to St. Vincent, the *Providence Mark* sailed between the islands for many years, managed by Mark Hall's family after his death. When Hurricane Flora hit Tobago in October 1963. The *Providence Mark* was caught at anchor there, and was completely destroyed.

On his boats, Mark Hall always wore "captain's whites" including a captain's hat. He would exchange this for a golfing cap when wearing ordinary clothes. In 1944 he was in his early 40s, and in that year he married Adella Lendore, who had been born in Venezuela but of Carriacou stock. Adella worked with Mark Hall on the boat "like a man".

The marriage was not blessed by children, but he had two children in Trinidad. Adella's sister, Drucilla is the mother of the musician Aiden Slinger.

Mark Hall had the ambition to see Europe, and this he did at age 50, but not before celebrating his half-century with a big party on the *Providence Mark*. While he was in Europe, Mark Hall began to feel ill. Shortly after returning to Grenada he had to seek medical attention, going to Barbados for further treatment, where he died at age 51. The Church of England in Grenville, where he lived for most of his life, could not hold the massive turn-out of people for his funeral service.

16. Henry, Vivian. Some of the beautiful furniture he crafted still adorns Grenadian homes. He started out managing Hubbard's Hardware, and then opened his own general store in St. George's situated on the corner of Halifax and Hillsborough Street. Besides furniture and other merchandise, his store sold fine cutlery and the best quality tableware.

17. Hosten, Mable. She was the sister of Lyle Hosten, one of Grenada's prominent lawyers. She married John Kerr.

I, J

18. James, Cosmos. He had a fairly large farm in St. David's and also owned a bus. He was the brother-in-law to Chasley David. His three sons were Alva, Curtis and Claude. Curtis died as a teenager, and Alva and Claude migrated to the United States, and worked in banking and computers (IBM) respectively. Both sons returned to retire in Grenada.

19. Johnson, Marjorie. Marjorie taught at Wesley Hall School, and later was transferred to the civil service. She married Dillon Baptiste, headmaster of the GBSS and later principal of the Church of England High School for Girls. On her retirement from the Public Service, Marjorie opened Lincarlene, a private primary school.

20 Jones, Franklyn. He was a clerk and lived at Old Fort.

K, L

21. Lambert, Olive. She was the sister of Dennis Lambert, one of Grenada's prominent lawyers who was a veteran of World War II. She was young, pretty and vivacious, and would have fitted perfectly with the crowd on the *Island Queen*. Nevertheless, she made the journey to St. Vincent on the boat that did arrive.

M

22. Marrast, Lucy. Lucy Marrast was a nurse who lived on Green Street. Some people remember that she has a lisp.

23. Marryshow, Ivor. Ivor was another of T.A. Marryshow's sons. He is described as being tall and lanky and a "gorgeous man". His mother worked at Evans' Store. He worked for F.M. Henry, a prominent lawyer. Ivor married Lucille Wilson from St. Paul's, the daughter of Joseph Wilson, a politician. Eventually, the couple migrated to the USA.

24. Mitchell, E.A. "Doc". "Doc" Mitchell owned and ran a restaurant at the corner of St. John's and Melville Streets and also had a cola factory. He lived above his restaurant, and had a son called Albert. A politician, he ran against T.A. Marryshow for the St. George's seat. He was the chaperon of the All Blacks Football Club.

N, O

25. Osborne, Phyllis. Originally from St. Lucia, she spent almost all of her adult life in Grenada. She worked at the Co-operative Bank almost since its inception in 1933, and until she was of very advanced age. Never married, it was said she was "married" to the Bank and to the Y.W.C.A. Phyllis was devoted to her church and gave voluntary service in the Church Office of the Church of England. She was the aunt of Marlene, Pat, Desmond, Vivian, and Denis Noel.

P, Q, R

26. Radix, Kenneth. Tall and lanky, he was a popular footballer and cricketer. He lived on the corner of Grenville and St. John's Streets. He became a pharmacist, and migrated to Venezuela.

27. Redhead, Gordon. Son of Dr. T.A. Redhead

28. Renwick, Derek. Although not of the medical profession, he was nevertheless called "Doc." He was the brother to Robin, Tony, Gordon, Ronald and Hugh. He studied agriculture at the Imperial College of Tropical Agriculture in Trinidad. On his return, he taught science at GBSS before eventually migrating to Canada. On the day of the excursion, he switched places at the last minute, giving his ticket to Ian Hughes, who badly wanted to travel with his girlfriend, Honey Rapier. This sacrifice was made much easier, as Derek had many friends on the *Providence Mark*.

29. Ross, Reginald. "Reggie" as he was called was well respected, and lived in Belmont. He is remembered as being short of stature. In later life, he became the manager of Everybody's Stores.

S

30. St. Bernard, Cosmo. Cosmo became one of Grenada's most prominent lawyers. He served as Chairman of the Public Service Board of Appeal, Chairman of the Grenada Co-operative Bank and Chairman of the Grenada Building and Loan Association. He was awarded the C.B.E. for distinguished public service, and in addition was appointed a Queen's Counsel. Originally from Duquesne, St. Mark's, his father was Allan St. Bernard, a planter, who often wrote articles for *The West Indian* Newspaper. His sister, Gloria, was the Secretary of the Grenada National Party led by Herbert Blaize, and one of the Mayors of the Town of St. George. His other sister was Yolande Bain. Reginald and Joslyn St. Bernard were his half-siblings. Cosmo St. Bernard was a member of the All Blacks Football Club.

31. Samuel, Cecil. He was about 35, at the time of the excursion, and an older supporter of the All Blacks Football Club. He is remembered for his magnificent Triumph motorcycle. He worked in the P.W.D. as a road officer. He married Maize Banfield and they migrated to the United Kingdom. Cecil returned to Grenada on his retirement.

32. Seales, Carden. He was a resident of Mt. Moritz

33. Shillingford, Edward. C. He was originally from Dominica. He had badly wanted to travel on the *Island Queen*, but had to settle for the *Providence Mark*. His brother, Hesketh, was the father of Claudia Joseph, Barbara Ann Shillingford, Geraldina Perotte, Alex Hood, and several other children. Edward worked at a lime juice factory on the wharf which Hesketh operated.

34. Stephenson, Lyle. He was from St. George's and had a very tall brother called Carlyle. He worked at *The West Indian* Newspaper, and stayed in that workplace until his retirement. In his youth he was an outstanding footballer, and scored many goals for the teams he played for — from primary and secondary school teams to adult clubs. He was a member of the All Black's Football Club,

35. Stroude, Alfonze. Alfonze Stroude lived on the Carenage and worked in the Government Printery. Later in life he owned and ran the Palm Grove Guest House with two of his sons.

T

36. Taylor, Fitz. He worked with the electricity company, and was a founder-member of the Technical and Allied Workers Union.

37. Toussaint, Neina. This lady lived on Green Street and later on Lucas Street. She was a dressmaker and eventually migrated to Trinidad.

U, V, W

38. Williamson, George. He was brother to Ernest, who was lost on the *Island Queen*, and the son of the Chairman of the St. George's District Board, Arnold Williamson, and Constance Williamson. Both George and his brother Ernest were graduates of GBSS. George studied bookkeeping and was slated to take over the office management of his father's business while his brother Ernest was to take over the garage. Instead he married a Miss Joyce DaBreo and migrated to the United States. Alice McIntyre is his sister.

APPENDIX III

The *Island Queen* Disaster

by Cosmo St. Bernard

The text of an article that appeared in the Grenadian Voice newspaper Friday, 30th July, 1999.

For many years, I had kept saying that one of these years I should write an account of my reminiscences of the *Island Queen* disaster, and now that yet another anniversary date will soon arrive, I have decided not to postpone this writing any longer.

The organiser/promoter of the August weekend excursion to St. Vincent for the August holiday weekend, which was to commence on Saturday 5th August, 1944, was a young man named Gordon Campbell, who hailed from St. Andrew's but lived mostly in the town of Saint George's where he was employed.

Gordon Campbell frequently organised and promoted dances, concerts, other entertainments etc., but his ventures were never successful, yet he was keen and continued to promote and organise similar activities.

About six weeks before the beginning of August 1944, Campbell conceived the idea of promoting an August weekend excursion to St. Vincent.

The All Blacks Club, which mostly played football and comprised principally old boys of Grenada Boys Secondary School, then the only Boys Secondary School in Grenada, and most of whom, including your humble servant, were employed in the Civil Service as junior clerks, was approached by Campbell to go on the excursion, and very early we the All Blacks Club decided that we would make the trip and made the necessary arrangements with Gordon Campbell.

After a few weeks, Campbell came back to us and said the way things had turned out, that his original idea of a one-boat excursion, the original single boat being the *Island Queen*, was no longer feasible.

The public response was so good, so many people wanted to go, that it became necessary for him to get another boat, and he had arranged to use the schooner *Providence Mark*, owned and captained by an experienced captain, Mark Hall, who operated principally between Grenville and Trinidad, but was not very well known in St. George's.

He had said that lots of young ladies in particular became interested in the excursion. They were mostly from St. George's and knew the well-known and well-liked Salhab family, the owners of *Island Queen*, and they definitely preferred to go on the *Island Queen*.

And we, the All Blacks boys, readily agreed at his request that we did not have any objections to being relegated to travelling by the second string boat, the *Providence Mark*, and thank God we did this and made this choice, and in fact travelled by the *Providence Mark*.

All the would-be excursionists assembled at the St. George's Pier on the Saturday afternoon, and after the customs formalities were completed, the *Island Queen* with her complement of about sixty (60) persons in all, I seem to recall, including the organiser Gordon Campbell, was the first to move away from the Pier, and the *Providence Mark* quite soon moved off and followed. Meanwhile, relatives, friends and well-wishers on the Pier were smiling and waving goodbyes to the passengers on the departing boats. Little did they or any of us know that so many of us would have been leaving for the last time and never to return.

From the beginning, the *Island Queen* was pursuing a more outward course, while the *Providence Mark* continued to hug closer to the coastline.

First, the General Hospital Point was rounded, with the two boats not maintaining any considerable distance between them, only that the *Island Queen*, which remained slightly ahead maintained a more outward course, and this continued as the boats progressed northwards along the Western Coast.

I might say at this stage that I have been travelling regularly by bus along the Western Main road, which is for the most part close to the coastline, from my earliest school days attending school in St. George's and was very familiar with the bays, inlets, districts etc. along the coast from St. George's right up to Duquesne Bay.

The boats progressed upwards from Moliniere, Happy Hill area, Brizan, Woodford, Black Bay, Concord, Grand Roy, La Poterie with the *Island Queen* a little ahead and consistently maintaining a much more outward course, and progressively moving further and further out and away from land and widening the gap, and by the time we got to Palmiste and then the town of Gouyave, the *Island Queen* was maintaining a course very much further away from the *Providence Mark*. Then Maran and the town of Victoria were reached and passed.

By the time darkness set in, we were still aware of the presence and position of the *Island Queen*, as by then there was a light up on a mast which could be clearly seen.

Both boats continued to make progress northward along the Western Coast, and by then the weather had become blustery and continued to be rainy, and my last recollection of seeing the *Island Queen* was when we were close to Duquesne Bay.

The *Island Queen* could be seen with its light on a mast very far away and appeared to be on the horizon literally as far as we could see, away from the coast, and I repeat that was the last place I recall having seen the *Island Queen*.

I cannot be sure about this but I have an idea that the time was then perhaps about 8:00 – 8:30 p.m.

The *Providence Mark* continued on and we arrived at Kingstown St. Vincent perhaps about 8:15 to 8:30 on the Sunday morning, and we soon realised that the *Island Queen* was not yet in port and we all began to feel good that we had beaten the "Queen" to it.

We were cleared by Customs, and by 9 a.m. or a little later the general feeling was that the *Island Queen* should by then have shown up, and when it had not by close to 10 a.m., cable and/or telephone calls began to be made to Carriacou, enquiring whether the *Island Queen*, which had engine and other troubles before, had got to Carriacou.

By midday on the Sunday, with concerns mounting, more calls began to be made to Grenada, Trinidad etc., and sea and air searches began to be made with no success.

The non-appearance of the *Island Queen* and the other excursionists threw a damper on the entire holiday outing, and this got progressively worse and gloomier as the hours passed. There was the Sunday, then the Monday and the final day, the Tuesday.

In spite of searches, Fleet Air Arm aircraft based in Trinidad, and all who became involved, the reports were consistent that there was no sign of the boat nor was wreckage, debris or flotsam ever seen.

The *Providence Mark* with its complement finally left St. Vincent late on the Tuesday evening and had an uneventful return trip to Grenada.

When the *Providence Mark* berthed on the Carenage opposite the First Station on Wednesday morning, there were crowds of relatives and friends lined up with great anxiety and gloom on their faces who had come to see who had really returned, as it was known that there were some last-minute switches between the two boats of a few of the passengers.

Those of us who returned got the feeling that a lot of the anxious relatives and friends, who had assembled on the Carenage, felt that they would likely never see again all who were not among those who had returned by the *Providence Mark* that morning. There are all sorts of theories and speculations as to what really happened to the I*sland Queen*, but it appears that it has been and will continue to remain one of the unsolved tragedies of the sea.

Cosmo St. Bernard

30th July, 1999

INDEX

A

Achilles, Albrecht 87, 88, 89
Adams, Edgar 97, 101, 150
Adams, Randolph 98
Adolf Hitler 79, 80, 3
Alexander, Douglas Gordon 40
Alexander, Jerome 179
Alexis, Carmen 34
All Blacks Football Club 134, 138, 180, 250, 251
Alleyne, Septimus 178, 181
Andrews, Bertie 31
Anglo-American Caribbean Agreement 119
Anglo-American Caribbean Commission 219
Antigua 74, 75
Antsley, A.H. 169
Archer, DeVere 135
Archer, Frieda 134
Archer, Shirley 134, 149, 152, 155, 173
Arthur, Bertram 158, 188
Arthur, Jackson Dunbar 44, 46
Aruba 67, 84, 112, 117, 85, 86, 209, 98, 176, , , 63, 56, 60
Auffermann, Hans 95, 96
Austruther, Fr. Godfrey 61

B

Bailey, Joe 29
Bain, Arthur 93, 102, 124, 144
Bain, Beatrice 34
Bain, Bert 4
Bain, Enid 61
Bain, Janice 102, 30, 62, 4
Bain, Kathleen 4, 34
Bain, Margaret 154
Bain, Michael 40
Bain, Ronald Wells 42
Banfield, George 195
Baptiste, Jack 199, 133, 204, 25, 36
Barbados 153, 154, 161, 169, 175, 194, 86, 89, 206, 91, 95, 96, 98, 30, 114, 115, 117, 41, 119, 120, 124, 125, 50, 7, 17
Barbados Advocate, the 194
Barker-Hahlo, George 17, 29
Barker-Hahlo, John 16
Barker-Hahlo, Rosamond 206, 16
Bartholomew, Claude 42
Bascus, Leah 40
Bauer, Ernst 98
BBC 2, 10, 49, 51, 52, 53, 54, 60
broadcasts 3
Beane Field 74
Berkley, Leonard 219

Bertrand, Mab 167
Birchgrove 206, 44, 61, 33
Bishop, Alimenta 112, 60
Bishop, Rupert 60
Blaize, Herbert 60
Blencowe, George 151
Blencowe, Louie Fraser 175
Bocas, The 107
Bonaparte, Basil 174
Bonaparte, Joan 175
Brathwaite, Godwin 38, 39
Brathwaite, Gordon 219
Brathwaite, James 18
Brathwaite, Roslyn 180
Brathwaite, Sam 180
Briercliffe, Sir Rupert 135
Bristol, Carol 89, 158
Buckmire, Shirley 145, 16
Butler, Tubal Uriah "Buzz" 208, , 7
Byer, Maurice 219

C

Calendar, Molly 219
Callendar, Oswald 169
Cape, Cosmos 111, 54, 42, 3
Carberry, Justice J.E.D. 169
Carlisle Bay 95
Carlsen Field 72
Carriacou 111, 116, 125, 139, 149, 150, 154, 186, 187, 188, 77, 211, 213, 102, 103, 104, 105, 107, , , 25, 30, 38, 39, 49, 16, 17
Cave, Joy 147
Chaguaramas 129, 69, 70, 71, 72, 73, 76, 88, 57
Chamberlain, Neville 80, 13
Charles, Bertie 180
Charles, Beryl 180
Charles, Monica 147, 178
Charles, Robert 114
Charles, Rosemary 180
Chin, Isaac 161
Christopher, Henry 42
Churchill, Winston 66
Church of England High School for Girls 137, 156, 167, 172, 91, 24, 30, 31, 33
Cipriani, Arthur Andrew 7
Clark, Kelvin Fr. O.P. 3
Clarkson, Thomas 8
Coard, Alfred C. 160
Cochrane, Dr. Edgar 142
Comissiong, Elma 38
Comissiong, Terrence B. 76, 195

Compton, Patrick 211
Compton, Ville 212
Coolidge Airfield 74
Cromwell, Leo 176, 93, 38
Cross, Ulric 9
Cruickshank, Arnold 75, 26, 3
Cruickshank, Kathleen 134, 138, 151, 181
Cruickshank, Michael Anthony 44
Curaçao 117, 66, 67, 75, 84, 86, 99, 60, 154, 165, , 56

D

DaBreo, John 76
Daggerrock 70
DaSilva, Pedrito 49
Date, Adrian 137, 150
David, Ivan 178, 215
David, Ronald Ivan 45
Davis, Josephine 26, 33
Declaration of Panama 66
DeDier, Ruby 94, 139, 185
de Freitas, Oona 137
de Freitas, S.G. "Papa" 98
deGale, Leo 40
deGale, Olga 40, 46
DeRiggs, Anthony 176, 182
DeRiggs, Hyacinth 133, 182
DeRiggs, Lucy 133, 173
de Riggs, Muriel 43
DeSouza, Rosie 181
Docksite 71
Dominica 186, 154, 161, 170, 30, 43, 45, 17
Donovan, Ethel Mary 38
Donovan, Nellie 76, 104, 108
Donovan, Selby 181
Dowe, Reginald 93

E

Edinburgh Field 72
Edwards, Cecil 106, 109, 77, 112, 38, 125, 50
Enoe, Headley 212
Evans, Terry 40

F

Ferguson, Lionel 29
Ferris, Joseph Sgt. 44, 46
Fleming, John 35
Fleming, John F. 142
Fletcher, Albie 43
Fletcher, Ena 135
Fort Reid 72

Francis, S.A. 182
Fraser, Alex 175
Fraser, Louie 137, 151

G

Gairy, Eric Matthew , 60
Garraway, Estelle 156, 18
Gentle, Alan 35
Gentle, Allan 42
Gentle, Eileen 137
Gentle, Gordon 52
George F. Huggins 237
Gibbs, Joseph 33
Gibbs, Kathleen 18
Gibbs, Mable 18
Gittens, Edna 172
Glasgow, Justina 178
Glasgow, Tina 135
Glean, Alister 52
Glean, Eric 17, 25
Gomas, Clythe 133
Gomas, Eileen 133
Gomas, Ena 133
Gomas, Erma 133
Gomas, Rita 133
Gomas, Sylvia 133
Gouyave 206, 105, 29, 42, 52, 63, 154, 4, 24, 3
Graham, Samuel 3
Grant, Jackie 101
Graves, George 40
Greasely, Hermione 4
Greasley, Hermione 24, 52, 63
Greasley, Sylvia 52
Grenada Boys Secondary School 134, 135, 101, 28, 41, 136
Grenada Volunteer Force 5
Grenville 90, 207, 103, 105, 109, 125, 133, 141, 153, 170, 185, 194, 29, 46, 50, 51, 53, 61, , 3, 4, 27, 3
Griffith, Elton George 59
Griffith, Joseph 206
Grimble, Sir Arthur Francis 166, 6, 170
Grow More Food campaign 22, 57
Gulf of Paria 71, 71, 85, 48
Gun-Monroe, Clarence 156
Gun-Monroe, Sheila 156
Gun-Munro, Christine 16
Gun-Munroe, Henry 195
Gun-Munro, Margaret 40

H

Hagley, Ermintrude 148
Hall, Mark 105, 141, 189, 195
Hamburg-American Line 2
Harbin, Laurie 18
Harris, Cecil 109, 42, 41
Haydock, Gordon 40, 42
Haynes Bakery 20
Heape, William Leslie 142
Hill, Molly 174
Hillsborough 104, 187, 133, 213, 39
Hinds, Mottley 41
Howell, Lennox 133
Hudson, Noble 15
Hughes, Alister 149, 188, 59
Hughes, Austin 173, 200
Hughes, Captain Earle 39
Hughes, Clarice 134, 135
Hughes, Dawn 135, 151, 155
Hughes, Dudley 141
Hughes, Earle 135, 173
Hughes, Ewart 155
Hughes, Ian 135, 141, 148, 173, 182, 187
Hughes, Roy 42
Hughes, Russell 135, 176
Husbands, Huille 135
Hutchinson, Gordon 53
Huton, Michael 212

I

Ireland, Jocelyn 195

J

Jacobs, Wilfred 179
James, Eileen 137
John, Carlyle 91, 106, 124, 55, 61, 62, 158, 24, 27, 38, 52
Jonas Brown and Hubbard firm 145
Jones, Ben 42
Julien, Annette 167
Julien, Paula 164, 176, 184
Julien, Willan E. 176, 181
Junker, Ottoheinrich 129

K

Kapitsky, Ralph 128
Kent, Betty 40
Kent, Edward 16
Kent, George 52
Kent, Leonard 50

Kerr, Rita 40
King George VI 9
King, Joseph 139
Kingstown 97, 101, 136, 149, 151, 152, 158, 163, 175
Kirby, Earle 199
Knight, Derek 42
Knight, Gittens 152, 195, 19
Knight, Ivor 135, 148
Knight, Monica 94
Knight, Thelma 173, 185

L

Lady Angela 164
Lady Drake 92
Lady Hawkins 92, 114
Lady Nelson 88, 89, 90, 91, 108, 114
La Grenade, Dora 43
La Guerre, Anastasia 102, 174, 25, 28
Lalbeharrysingh, Cynthia 24
Lalbeharrysingh, Dolphus Gardner 24
Lalbeharrysingh, Lynda 174, 24
Land Lease Agreement 67, 70, 74
Lang, William Grahame 44
Lashley, Ethel 172
Lashley, Marcella 145, 207, 94, 172, 30
Lucas, C.H. 12
Lumsden, Esmai 182, 183
Lusan, David 195

M

MacLeish, Marjorie 137
Maggie Mitchell's Bakery 20
Mahy, Jane 38
Maitland, Adina 172, 209, 93, 3
Malins-Smith, Bert 206
Malins-Smith, Dennis 206, 39
Malone, Clement 51
Mancini, Keith 40
Mark, Randolph 38
Marryshow, Hans 133
Marryshow, Julian 44, 53
Marryshow, Louise 133, 172
Marryshow, Sheila 133, 176
Marryshow, T. A. 133, 142, 194, 163, 178, , 44, 36
Martin, Frieda 18
Matron Augustine 3
Mauricette, Rooney 42
McDermott Coard, Frederick 20

McIntyre, Alister 200, 204, 60, 167, 26, 4, 20
McIntyre, Colin 140, 148, 179, 195
McIntyre, Eileen 15, 26, 35
McIntyre, Meredith 34, 60
McIntyre, Mollie 18
McLawrence, Patrick 212
McLeish, Lennard 55
Medford, Clive 41
Middleton, Charles 8
Minors, Merle 137
Mitchell, Cylinford Augustus 28
Mitchell, Dora 56
Mitchell, Norris 4, 28
Monroe, Monica 41
Moore, Eileen 43
Moore, Elaine 43
More, Hannah 8
Morne Jaloux 34
Munich Agreement 80
Murray, Alma 148, 181

N

Nunez, Emeline 30

O

O'Brien Donovan, William 22
Ogden, Major David 69
Ogilvie, Harry 104, 52, 38
Ogilvie, Ralph 143
Ogilvie, Ralph Sr. 141
Operation Neuland 78, 82, 84
Osborne, Phyllis 180, 199
Otway, David 180, 17
Otway, George 153, 155
Otway, Rhona 204

P

Pantin, Margaret 155
Parke, Judith 105, 56, 62, 63, 142, 164, 23, 28, 52
Parris, Humphrey 138, 28, 174
Parris, Joan 28
Parris, Madeline 157, 177, 178
Paterson, Lucy 133
Paterson, Maurice 199
Patrice, Peter 212
Patterson, Claude 148
Patterson, Lucy 158
Payne, Herbert 108, 41
Pearls 75, 102, 104, 105, 207
Pearls Airfield 74, 75
Peebles, Margaret 109
Phillip, Beatrice 178

Phillip, George 45
Phillip, Margaret 109, 181, 75, 94, 147, 178, 183, 204, 62
Phillip, Perle "Jappy" Moore 133, 178
Phillips, Angela 102, 111
Phillips, C.A.O. 102
Piening, Captain Adolf 97
Pilgrim, Arthur 26, 34
Pilgrim, Evelyn 137, 34
Pilgrim, H. H. 179
Pilgrim, Hugh Henry 18
Pilgrim, Norma 137, 158, 182
Pitt, Bertrand 104, 105, 114, 4, 102, 53, 61
Pitt, Cyril S. L. 4
Popham, Sir Henry Bradshaw 142, 6
Preudhomme, Gordon 141
Prince, Beresford 195
Principles Underlying British War Time Propaganda 1939 13
Purcell, Clara 169

Q

Queen Elizabeth 9
Queen Victoria 8

R

Rapier, D E.W. 40
Rapier, George 141, 148, 158, 174, 179
Rapier, Helena 135, 148
Rapier, Hilda 38
Rapier, Julian 141, 188, 102, 38
Rapier, Rupert 53
Redhead, Irvin 109, 195
Redhead, Wilfred A. 186
Renwick, Charles F. P. 142, 5
Renwick, Derek 135, 182
Renwick, Florence 214
Renwick, Robin 180
Report of the West Indian Royal Commission 195, 196, 197, 198, , 218, 10
Richards, Joan 138, 167
Richards, Justice G.E.F. 161
Richards, Lester 138
Richenburg-Klinke, Kurt 98
Richmond Hill 76
Roberts, Ben 204
Rodriquez, Carlos 148
Roosevelt, Franklin 66
Rose Marie 153, 185, 154, 157, 57

Ross, Colin 44, 46
Ross, James 44
Ross's Point 76, 108
Rowley, Captain H.E.A. 153, 155
Rowley, Iris 170, 172, 177
Rowley, Redvers 103, 49, 167, 170, 172
Rowley, Robby 198, 103, 49, 39

S

Salhab, André 143, 195
Salhab, Chykra 132, 138, 139, 144, 145, 189, 190, 192
Salhab, Cynthia 43
Salhab, Dora 174
Salhab, Elinor 206, 63
Sardine's 20
Scoon, Doreen 133, 187
Scoon, Layinka 172
Scoon, Laynika 133
Scoon, Marie 172, 173
Scoon, Paul 63
Scoon, Winifred 172
Scott, Captain John 164
Searles, Arthur 41
Searles, Cogland 41
Shannon, Harold K. "Buzz" 44
Sharp, Granville 8
Shears, Dorabella 103
Shillingford, Ruby 43
Slinger, Dawn 137, 167
Slinger, Denise 137, 167
Slinger, Dr. Evelyn 137, 172
Slinger, Drucilla 141, 204
Slinger, Paul 89
Smith, Ray 219
Sobers, Geraldine 91, 62, 26
Springer, Marjorie 173
Springler, Vernon 106
St. Andrew 207, 104, 105, 144, 156, 27, 32, 52, 61, 4
St. Bernard, Cosmo 134, 148, 189, 213
St. Bernard, Jocelyn 133
St. Bernard, Joslyn 158
St. Bernard, Reginald 133, 158
St. David 110, 153, 75, 96, 33, 61
Steele, Clayton 192
Steele, Dunbar 176, 14
Steele, Henry 192
Steele, John Louis 193
Steele, Louie 109
Steele, Osborne 143, 180, 192, 195
Steele, Sydney 59
Steele, Thelma 174, 192

St. George's 75, 76, 91, 93, 94, 102, 103, 39, 41, 51, 132, 133, 136, 59, 139, 63, 141, 142, 2, 3, 4, 5, 6, 15, 16, 20, 21, 24, 25, 26, 29, 32, 33
St. George's Harbour 109, 110, 77, 1, 2, 91, 137, 138, 144, 164
St. John 133
St. Joseph's Convent 136, 137, 182, 204, 38, 158, 30, 31
St. Louis, Lincoln 163, 172, 176
St. Lucia 153, 154, 158, 161, 162, 170, 184, 199, 87, 89, 90, 91, 95, 43, 45, 123, , 17, 30, 35, 7
St. Patrick 28, 49, 52, 59, 147, 24
St. Patrick's Roman Catholic School 158, 38
St. Paul's 94, 133, 167, 26, 29, 63
St. Vincent 176, 178, 179, 180, 181, 184, 77, 196, 199, 96, 97, 98, 99, 111, 116, 118, 130, 131, 132, 133, 134, 135, 136, 137, 138, 141, 144, 145, 147, 148, 149, 150, 152, 153, 154, 155, 157, 159, 161, 164, 166, 167, 169, 170, 172, 173, 175, 201, 24, 30, 34, 43
St. Vincent Grammar School for Boys 136
Swap, Bernice 138
Swap, Gunny 205

T

Tanteen 76, 178, 77, 41, 46
Teka, Oris 55, 63, 173, 37
Telescope 105
Telescope beach 104
Telescope Point 144
The Times newspaper 161, 162
 report 132, 168
The Vincentian newspaper 161
 report 160, 165, 166
The West Indian newspaper 143, 161, 54, 55, 162, 164, 169
 report 140, 195, 159, 160, 169, 171, 53
Trinidad 120, 122, 123, 124, 125, 126, 128, 129, 138, 141, 144, 150, 153, 154, 156, 160, 164, 167, 169, 179, 181, 182, 67, 184, 185, 69, 186, 70, 71, 72, 73, 190, 74, 75, 76, 77, 78, 79, 84, 201, 85, 86, 87, 205, 206, 207, 91, 208, 209, 210, 211, 95, 97, 98, 99, 103, 105, 106, 107, 108, 111, 112, 115, 117, 119, 49, 54, 56, 57, 58, 59, 63, , , , 21, 27, 29, 30, 31, 33, 41, 42, 43, 44, 45, 46, 47, 48, 15, 16, 17
Trinidad Guardian newspaper 55
Trinidad Royal Navy Volunteer Reserve 122

U

Umtata 88, 89, 90
Union Island 187
Union Jack, the 8, 11

V

Venezuela 154, 156, 66, 67, 185, 190, 79, 99, 156, 165, 22, 56
Victoria 102, 105

W

Waller Field 72
Watts, Cecil Sitwell 33
Watts, John 182, 59, 33
Wells, Brenda 156, 93, 34
Wells, Captain Sydney 153, 155
Wells, Owen 50
Wells, Sydney 195
Whitaker, Charles 75, 76
Wickham, Viola 169
Wilberforce, William 8
Williamson, Arnold 15, 142, 6
Williamson, Constance 15
Williamson, Ernest 146
Williams, Vivian 40
Wilson, Ignatius 59
Windward 105, 116, 211, 213, 214
Woodruff, Cuthbert 144
Woodruffe, Myra 40
Wright, A. A. 6

Y

Yearwood, Jenny 94, 29

Z

zeppelin 124, 125, 72, 127, 62, 63, 154